THE FIGHT FOR THE
SOUTH DOWNS

First published 2024

Copyright © Robin Crane 2024

The right of Robin Crane to be identified as the author of this work has been asserted in accordance with the Copyright, Designs & Patents Act 1988.

All rights reserved. No part of this book may be reproduced, stored in a retrieval system, or transmitted in any form or by any means, digital, electronic, electrostatic, magnetic tape, mechanical, photocopying, recording or otherwise, without the written permission of the copyright holder.

Published under licence by Brown Dog Books and The Self-Publishing Partnership Ltd, 10b Greenway Farm, Bath Rd, Wick, nr. Bath BS30 5RL, UK

www.selfpublishingpartnership.co.uk

ISBN printed book: 978-1-83952-757-9
ISBN e-book: 978-1-83952-758-6

Cover painting 'Bad weather over Bury' by Gordon Rushmer
Cover design by Kevin Rylands
Internal design by Andrew Easton
Aerial photography courtesy of Ross Oliver @ Airscapes.co.uk

Printed and bound in the UK

This book is printed on FSC® certified paper

THE FIGHT FOR THE SOUTH DOWNS

The long struggle to protect one of Britain's most treasured landscapes

ROBIN CRANE

Edited by Margaret Paren

The South Downs National Park

CONTENTS

Author's Introduction		7
Dedication		9
Preface		10
Chapter 1	The South Downs and the Western Weald	13
Chapter 2	The Pioneers 1923-1966	28
Chapter 3	An Imperfect Hand	34
Chapter 4	The Formative Years 1990-1992	39
Chapter 5	Marking Time 1992-1997	53
Chapter 6	Change Afoot	61
Chapter 7	Progress	66
Chapter 8	The Breakthrough	83
Chapter 9	First Moves Towards Designation	91
Chapter 10	The Next Steps	99
Chapter 11	The Statutory Local Authority Consultation and the Designation Order	106
Chapter 12	The Inquiry: First Phase	114
Chapter 13	Major Setbacks	128
Chapter 14	The Fight Back	133
Chapter 15	The Inquiry Re-opens	143
Chapter 16	Victory at Last	160
Chapter 17	The Final Stages	169
Tribute by Len Clark		178
Acknowledgements		179
Final List of SDC Membership		180
Bibliography		185
Index		187

AUTHOR'S INTRODUCTION

My first encounter with the South Downs was in helping to haul a heavy trailer up the precipitous slope to Amberley Mount where our school signals section set up a camp for several days. Whilst carrying heavy wirelesses up and down these idyllic hills was not the ideal introduction, I was nevertheless overwhelmed by the breathtaking scenery and the flora and fauna. I had a passion for wildlife and, on another occasion, I visited Amberley Wildbrooks which has been renowned as a haven for naturalists for hundreds of years.

So, when my new career as a film-maker necessitated me moving back to the South of England in 1967, I instinctively chose Sussex, which was also known to have one of the best Wildlife Trusts in the country. Whilst the Trust's primary task at that time was the securing of the best sites as nature reserves, we were equally concerned about the frightening deterioration of the wider countryside through run-away farming policies, hedgerow destruction and the deadly effects of pesticides and pollution. In due course I was convinced that the creation of a national park would provide the strongest status for conservation available.

However, those organisations that joined the campaign to fight for a South Downs National Park reflected concerns far broader than nature conservation. The Campaign to Protect Rural England (CPRE) wanted to ensure the best planning protection, the Sussex Rural Community Council were passionate about making the countryside a better place for everyone to live or work in and enjoy. The Ramblers Association wished to promote access to the countryside, the archaeologists to protect the legacy of earthworks, ancient sites and artefacts. The Campaign for National Parks (CNP) provided much expertise and knowledge of existing national parks. Then there were many local residents in Sussex and Hampshire who loved their marvellous countryside and wished to ensure that it was in safe hands now and for future generations.

Over the years the South Downs Campaign evolved to meet the changing circumstances. Individuals came and went. I was extremely fortunate to be supported throughout the life of the organisation by exceptionally able and knowledgeable people and I always looked forward to meetings, however tough the going. Their work was of a high standard that stood up to many challenges. It was a huge privilege and joy for me to have had the opportunity to work with such an enthusiastic and delightful team.

When we started our Campaign, I did not anticipate the powerful resistance we would face including that from many politicians and even the Countryside Commission, then responsible for the designation process.

There were times when the South Downs Campaign had little to do and others of frenetic activity. At our inaugural meeting I was invited to chair it 'for the time being'. Twenty years later I still was chairman despite my offering to stand down on several occasions: none of us anticipated that we would have to sustain our campaign for that length of time.

All this work meant I could spare little time to pursue my nature conservation interests. I therefore confined my activities to studies of my local patch: the heathlands of Iping and Stedham Commons. These resulted in two papers based on 20 years of research: one on the Silver-studded Blue butterfly and the other on the bird population. It was rewarding and refreshing work far removed from the South Downs Campaign.

What kept me going through all the long years of the Campaign was the South Downs themselves. My filming career for the BBC and holidays with my wife took me to many exotic places, but to my mind nothing exceeds the very special qualities of the area that is now the South Downs National Park.

DEDICATION

This book is dedicated to all those who have fought to protect and enhance the South Downs and the Western Weald over many years, especially those supporters who sadly passed away during the course of the South Downs Campaign and did not live to see the fruits of their labours:

David Clegg

Nigel Paren

Denis Payne

Dr Francis Rose MBE

Sheila Schaffer

John Venning

Kath Worvell

A special mention must be made of our passionate campaigner and founder member Paul Millmore who fought for a national park for so many years. It was a tragedy that he died in 2012 so soon after the South Downs National Park was established. He surely would have continued to apply his considerable energy and expertise in supporting the National Park for many years ahead. It is a fitting memorial to his work that the library at the South Downs Centre in Midhurst is named after him.

PREFACE

This is the story of a twenty-year campaign to secure the highest level of protection for a swathe of countryside in south-east England stretching from Winchester in the west to Eastbourne in the east: a South Downs National Park.

The South Downs Campaign was formed when just seven people met at the Sussex Wildlife Trust headquarters in 1990. By the time the dream of a South Downs National Park was realised, the Campaign had swollen to a movement of 159 national, regional and local bodies. They ranged from conservation, recreation and amenity groups to local businesses and town and parish councils.

The story is told from the perspective of a number of those who were most heavily engaged in the Campaign. In telling their story I have drawn heavily on the welter of papers and meeting notes produced over the years as well as their contemporary diary entries and memories. I have not attempted to incorporate the memories of all who shared our vision and contributed so much but were not so directly involved. Nor have I tried to describe the story from the perspective of those who opposed us, sometimes vociferously. I have endeavoured to be fair in reflecting some of their reasoning and have not cast doubt on their commitment to a better deal for the Downs. We shared a common goal but differed profoundly on how that might best be achieved.

The first chapter describes the area that became the South Downs National Park with the purpose of helping the reader to understand why so many people feel such a commitment to it. I have then devoted two chapters to a foreshortened description of the earlier attempts to protect the area, to establish national parks in England and Wales and to enhance the countryside. I have also included something about the earlier endeavours of two of our founding members to achieve that same end, sufficient I hope to provide the reader with enough context to appreciate the main story, that of the South Downs Campaign.

PREFACE

© *Kris Pawlowski*

Key figures in the creation of the South Downs National Park at the unveiling of a plaque in 2020 to mark the signing of the Order to create the South Downs National Park in 2009

From left to right Chris Todd, Hilary Benn, Margaret Paren and Robin Crane

The Campaign's logo

CHAPTER ONE

THE SOUTH DOWNS AND THE WESTERN WEALD

Seven Sisters
© *Steve Oldfield Sussex Wildlife Trust*

"It would be difficult to find anywhere in the world an area of comparable size which exhibits so perfectly the responses of plant, animal and human life to the stimuli of varied physical environments as the Weald, which Londoners have on their doorstep." New Naturalist "The Weald"

(Note: Some writers, such as the authors of "The Weald", include the chalk of the North and South Downs as being part of the Weald. In this book I use the standard definition of the Weald, that is the older strata of mainly sandstone and clay that lie between the chalk hills of the North and South Downs)

Cutting a swathe through three counties of south-east England, the South Downs National Park stretches from Eastbourne in the east to Winchester in the west. In the east the National Park meets the sea at the iconic Seven Sisters: dazzling white cliffs of chalk, the highest in England. Inland, the chalk hills, Kipling's 'blunt, bow-headed whale-backed downs', are open and treeless. Further west, though more wooded, 'these majestic mountains' as Gilbert White called them, always retain an open aspect with panoramic views of the sea to the south and wide vistas of southern England from the top of the chalk escarpment looking north. As well as these chalk hills, the National Park encompasses the Western Weald, a mysterious world of sandstone ridges, clay vales and wild heathland: one of the most anciently wooded parts of England.

The character of all countryside is ultimately determined by the underlying rocks which shape the landscape and dictate the nature of its vegetation, wildlife and human occupation. This is particularly relevant in this very special part of south-east England where the surface geology was not disturbed by the glaciations of the last Ice Age. The magnificent hills of the South Downs, which dominate the region, consist almost entirely of chalk, an exceptionally pure, soft, porous limestone formed from marine organisms. Their structure is clear for all to see at Beachy Head and the Seven Sisters where the dramatic white cliffs reveal the underlying structure of the chalk with its layers of flint. In contrast, the area which became known as the Western Weald has an unusual variety of different geological features within a relatively confined area and these have evolved into a series of distinct landscapes. They range from sandy heathlands to firmer sandstone ridges and the Wealden clay vales which are so wet in winter and rock-hard in summer that they were the last area in England to be colonised by man. The whole is drained by a series of seven rivers, each with its own unique character.

The most unforgettable characteristic of the South Downs is their extraordinarily elegant form, epitomized by Mount Caburn near Lewes. There are no jagged rocks. Every feature is gently rounded and pleasing to the eye. They have always held a special place in the hearts of those who know them. Many writers and naturalists have been seduced by their unique beauty. To mention just a few: Rudyard Kipling, Hilaire Belloc, Edward Thomas, W H Hudson and Gilbert White, all conveyed the special attributes of this quintessential English landscape in their works.

The steep north-facing scarp slopes of the Downs stand high above the surrounding landscape and are visible from far away. One can drive all the way

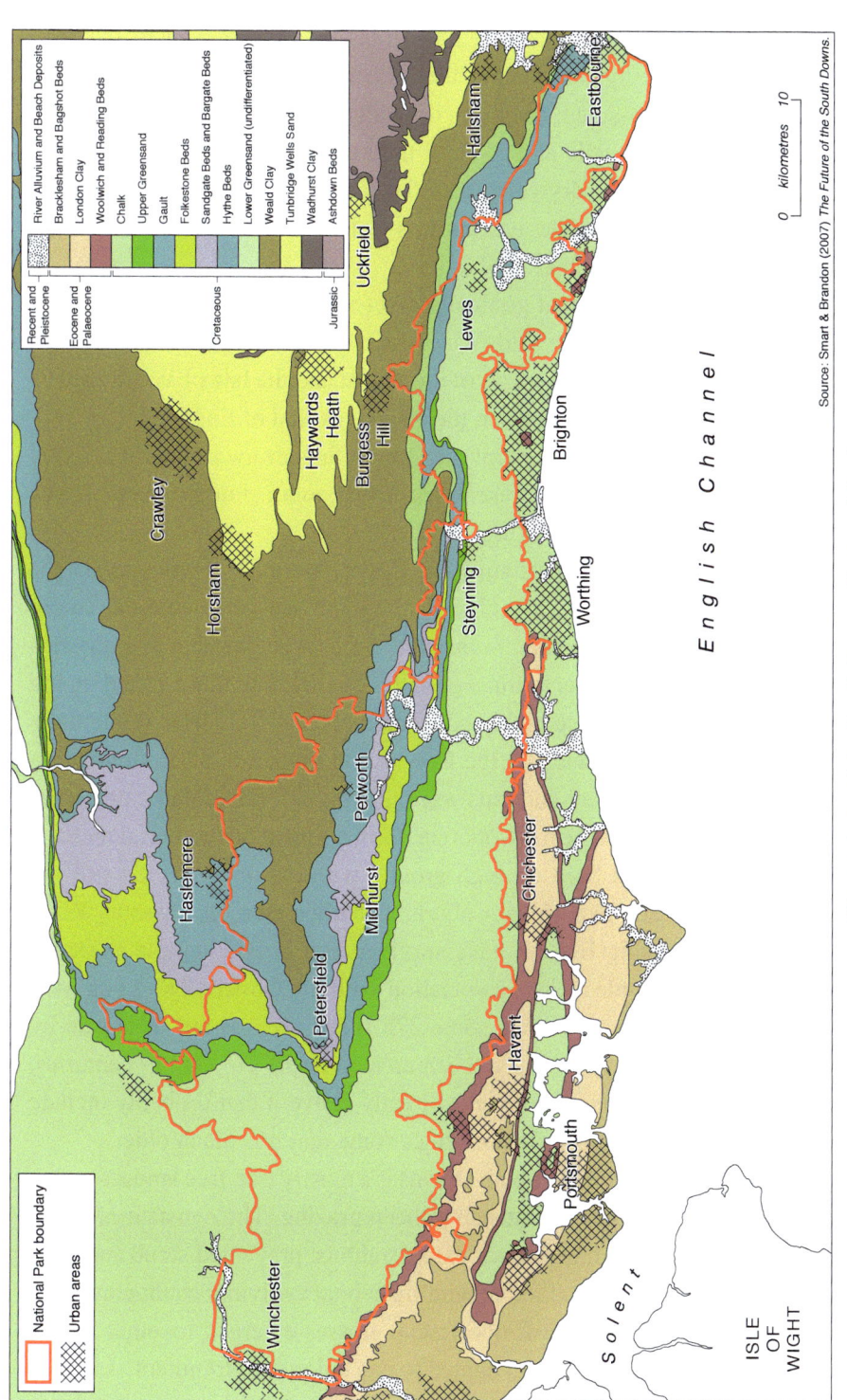

The geology of the South Downs National Park

from Eastbourne to Petersfield, some fifty miles (eighty kms), with the bold South Downs perpetually dominating the scenery. They are intersected by a series of river and steep-sided, deep-cut, dry valleys that can lead people into a world remote from the bustle and noise of the nearby roads or seaside resorts. The most spectacular of the dry valleys are the celebrated Devil's Dyke north of Brighton and the equally impressive, but less well known, Rake Bottom on the side of Butser Hill in Hampshire.

Further west the escarpment gradually gives way to more extensive rolling countryside with wide vistas: Old Winchester Hill, a formidable Iron Age fortress, provides breath-taking views of the Solent and the Isle of Wight; and St Catherine's Hill, views of Winchester, the ancient capital of England.

In Sussex, the gentler dip slope of the Downs falls away gradually towards the coastal towns and the ever-present sea to the south. Further west, it lies further inland and is more undulating.

With their light, well drained soils the Downs were an obvious target for early settlers. Flints found on the surface or by the seashore were extremely important for the hunter gatherers. In the New Stone Age, farmers dug quarries and mine shafts to gain access to flints of better quality than those found on the ground. From these they made highly polished tools that were used to facilitate the clearance of the woodlands on the Downs and the successful introduction of agriculture. Wheat, barley and oats were grown, and sheep and cattle were grazed in grassy areas. Many earthworks such as the flint quarries, burial mounds and a string of hill forts built on high ground with dominating views of the surrounding countryside like Mount Caburn, Ditchling Beacon, Cissbury Ring, The Trundle and Harting Beacon have survived and provide ample evidence of these prehistoric peoples' close association with the Downs. The Long Man of Wilmington is a massive edifice some 275 ft (72 metres) tall. It is said by historians and archaeologists to have been created in the 16th or 17th centuries, but the presence of complex earthworks directly above it that probably include prehistoric long and round barrows convince some that it is much older.

Today the eastern end of the South Downs is a mostly tree-free landscape that for centuries had been used for intensive sheep grazing. The constant nibbling by vast flocks of sheep and, to a lesser extent, rabbits, prevented scrub and trees from growing and led to a rare kind of wildlife habitat evolving: chalk grassland. On the well-drained springy turf, fine fescue grasses emerged together with a wonderfully rich carpet of flowers adapted to this unique environment. As many as 40 different species of flowering plants have been recorded in a square metre.

These plants attract a wealth of insects: butterflies, bees, grasshoppers and more, some of which are wholly dependent upon the chalk flora. Sadly, changes in agricultural practices have led to the loss of most of this habitat but there are still some stunning areas that have either been conserved by sympathetic farmers and landowners or secured as nature reserves. Despite the substantial loss of the old 'unimproved' chalk turf, a wonderful array of wildflowers and insects still linger on footpaths and bridleways and add to the pleasure of visitors. The 100 mile (160 km) long South Downs Way follows the old routes along the chalk escarpment and ridges for the whole length of the South Downs from Eastbourne to Winchester and is the most used National Trail in the country.

The South Downs in West Sussex completely change character west of the river Arun, partly because much of the chalk here is overlaid with superficial layers of clay-with-flints. The Downs become much more wooded. The ancient woodlands on the sheer northern slopes stretch almost unbroken to the Hampshire border, some of which are believed to have been unaltered since Neolithic times. No similar woods are found in the other chalk regions of Britain. More accessible for walkers are the very attractive beech woodlands of comparatively recent origin, that provide spectacular autumn colours. The magnificent ancient yew wood at Kingley Vale National Nature Reserve near Chichester is regarded as one of the finest of its kind in Europe. Much of the reserve is covered in the dense canopy of centuries-old yew trees, including weirdly shaped giants with gnarled branches and often multiple trunks.

The naturalist W.H.Hudson wrote *'during the whole fifty three mile length from Beachy Head to Harting the ground never arises above the height of 850 feet, but we feel on top of the world.'* Indeed, there are many breathtaking viewpoints along the Downs. One of the most noteworthy is the National Trust's Harting Down which provides one of the best places to cast one's eye over the relatively undeveloped and undulating Western Weald to the north. Here one can see the succession of sandstone ridges intersected by clay valleys that eventually rise to Black Down, at 920 feet (280 metres), the highest point in Sussex. On these Greensands are to be found a sprinkling of heathland interspersed with a patchwork of fields, hedges, copses and remnant ancient woodland.

The East Hampshire Hangers, which begin immediately west of Petersfield, are another dramatic geological feature. Their east facing slopes form the edge of the Hampshire chalk and Upper Greensand, the two types of rock separated by an ever-widening spring-line terrace. Clasping precariously to their precipitous slopes, ancient woodland, thought to pre-date the Ice Age, predominates in

The wooded north-facing scarp slope of the South Downs at Poynings
© Airscapes.co.uk

A view from Hampshire Hangers
© *Mischa Haller South Downs National Park Authority*

marked contrast to the pastoral fields and orchards of the terrace. Northwards of Petersfield lies one of the most famous villages of them all, Selborne. This was the parish of the Reverend Gilbert White, whose book 'The Natural History and Antiquities of Selborne' was first published in 1789 and has become the fourth most published book in the English language. A few miles further north lies Chawton, for a long time the home of Jane Austen.

Human influence on this Wealden landscape began further back in time than once was generally believed. There is growing evidence that Mesolithic hunter-gatherers, perhaps as early as 8000 years ago, deliberately used fire to create open areas in the forest, either to make hunting easier or to promote the growth of hazel, whose nuts were a key element in their diet. The impact of this human activity is most evident on the fragmented areas of heathland in the Western Weald on the Lower Greensands. Carpeted with purple heather in late summer, they may appear to be wild and untamed but are far from natural. They are the product of centuries of active management by man gathering wood, gorse, bracken, turf, peat and heather or burning the heath to benefit their grazing livestock. Once these ancient practices cease, the heathlands rapidly become dominated by birch and pine woodland, bracken and coarse grasses. The heather and its associated fauna gradually disappear. This highly specialized habitat is confined to Western Europe. Tragically 95% has been lost to development, forestry or neglect in the past 150 years. Southern England contains the largest and highest quality remaining tracts of this precious habitat.

Nearly all the surviving heathlands of the Western Weald are now actively managed to retain their characteristic plants and animals. There are some rare and important bird species such as the Dartford Warbler, Stonechat, Nightjar and Woodlark and a large number of invertebrates, including dragonflies, spiders, bees, wasps, flies and butterflies that are rare and confined to these specialised wet and dry sandy acid habitats. Woolmer Forest in East Hampshire was once one of Gilbert White's favourite haunts. This is the largest area of lowland heath in England outside the New Forest. It is the only place in Britain of this scale where all twelve of Britain's native amphibians and reptiles are found. Despite being an Army training area, today it is actively managed for wildlife.

These heathlands provide superb opportunities for those seeking access to open country. In many places it is possible to walk or ride quite long distances surrounded only by heather, birch and Scots Pine, with glimpses of the South Downs in the distance. There is a real sense of wildness. From viewpoints such as Older Hill, Iping Common or Longmoor Inclosure there are magnificent panoramic

Iping Common, a typical West Sussex heathland
© *Robin Crane*

Sussex cattle grazing on Ebernoe Common wood pasture
© *Mark Monk Terry Sussex Wildlife Trust*

views of semi-natural wooded countryside. The Serpent Trail is a 64-mile (103 kms) long footpath linking Haslemere to Petersfield and Midhurst that has been established to join up the remaining heathland areas of the Western Weald.

The heavy land of the unyielding Wealden clay vales is distinctive for its thin strips of ancient woodlands and hedgerows known as 'Shaws' or 'Rews' that divide many of the small fields – a landscape that has changed little since mediaeval times. Many splendid oak trees line the road verges and stand majestically in fields. The whole area remains one of the most wooded in lowland Britain with over half defined as ancient woodland.

Two woods are of special interest. Some parts of 'The Mens' north of Fittleworth have been there for at least a thousand years. This extensive wood, on greensands and clay, had not been managed for over a hundred years when the Sussex Wildlife Trust acquired it in the 1960s. It is a remnant of the once dense forest 'Andredswald'[1], that was described by the Venerable Bede in AD731 as *'thick and impenetrable and the haunt of large herds of deer and swine.'* Today 'The Mens' is as near as one can get to a natural forest with a variety of trees in different stages of growth and decay. The reserve is renowned for its bats and fungal flora.

For several years Edward Elgar lived at Brinkwells, an isolated cottage close to The Mens, where he wrote his cello concerto. Whilst there he also wrote a beguiling piano quintet which has been linked to a legend in which a group of dead, twisted trees nearby were said to be in the forms of Spanish monks struck by lightning.

A few miles north-west of The Mens is Ebernoe Common nature reserve. This oak and beech common is an ancient wood pasture where grazing has now been re-introduced. The wood is exceptionally rich in wildlife and is regarded as one of the finest woodlands in southern England. A furnace pond within the reserve provides links to the once extensive 16th Century Wealden iron industry and there is an 18th Century brick kiln that used local clay and firewood until the 1930s.

The richness and variety of the Downs and the Western Weald is enhanced by seven rivers. Chalk is porous and acts as a sponge and a filter, so crystal-clear water is released from springs. Two of the finest chalk rivers in England are the Hampshire Itchen and its smaller eastern neighbour the Meon, the latter flowing through the two picturesque villages of East Meon and West Meon.

---

1 Andredswald was the term used by the Anglo-Saxons to describe the great forest lying between the chalk hill of the North and South Downs.

The River Meon
© *Charlie Hellewell South Downs National Park Authority*

Fallow deer at Petworth Park
© *The National Trust photo Tom Cook*

The Itchen, famous for its game fishing, has been granted the highest level of international protection for its flora and fauna. Scattered along the river valley are a number of lovely historic villages: Itchen Stoke, Itchen Abbas, Martyr Worthy and Abbots Worthy to name but a few. It is also well served by a network of footpaths and bridleways, including the Itchen Way, Kings Way, Monarchs Way and Wayfarers Way.

The River Rother rises on the chalk just south of Selborne. It is the only river in the area that does not cut through the Downs. It flows through the Western Weald first southwards before turning sharply eastwards at Petersfield. It has a complex hydrography with tributaries joining it from both the chalk and acidic Greensand before, after 30 miles (48 kms), it joins the river Arun at Pulborough. A meandering, sand-bottomed river within a well-wooded valley, it is regarded as one of the least altered rivers in southern England and is the only river entirely encompassed within the boundary of the National Park.

Further east the Arun, Adur and Ouse all rise to the north and dissect the South Downs on their way to the sea. All are tidal on their lower reaches. All are impressive, with their own distinctive character. Perhaps the best known is the Arun. Directly to the north of the Downs the alluvial grazing marshes and drainage ditches of the Amberley Wild Brooks and Pulborough Brooks nature reserves provide a haven for rare plants and dragonflies and wintering waterfowl. As the river emerges from its downland valley to the south the famous town of Arundel, with its castle and cathedral, dominate its passage to the sea. The most easterly of all the rivers is the Cuckmere. A smaller river best known for the views of its meanders, wet meadows and shingle beach from Seaford Head or the Seven Sisters: a river mouth uncluttered by commercial development.

From earliest times there have been strong and enduring human links between the contrasting terrain of the Downs and the Weald. Those ancient peoples who first inhabited the chalk Downs soon began to exploit new areas on the more varied soils of the Weald. In the river valleys the wetter pastures were grazed in spring and hay was cut in summer. Wood pastures were created where cattle could graze amongst the trees of the forests, and pigs could forage. Woodlanders felled mature trees for constructing buildings, and shrubs such as hazel were coppiced in rotation to produce fencing, firewood and wattle for lining houses. Ancient droveways, that provide excellent opportunities for walkers today, were created for the movement of people and their livestock from the chalk grasslands to the fields and wood pastures of the Weald. It is no accident that in West Sussex many of the parish boundaries lie south to north across the various terrains of chalk, sand and clay. The synergy between the Downs and Weald are well represented at the Weald and Downland Living Museum located at Singleton, north of Chichester. Here, examples of vernacular buildings from across the south-east of England can be explored in the stunning setting of the West Dean Estate.

In Mediaeval times the church and the barons acquired large areas of this prosperous region of England. Wealthy landowners established enclosed deer parks, where they could hunt, undisturbed by the peasantry. The great estates such as Norfolk, Goodwood, Wiston, Glynde and Firle gradually evolved to combine farming, forestry and sporting activities. Today large areas are still under their careful stewardship and add richness to this magnificent countryside.

I have already named some of the many attractive villages to be found in the area. Unlike some other parts of England here there is no one landscape, no one vernacular, no one type of building material in use. All have their own character and history. To take just a few examples: Hambledon, surrounded by the sweeping landscape of the Hampshire Downs; the home of cricket before the opening of the Lord's Cricket Ground and an early location for the renaissance of the English wine industry. Amberley, a picture postcard village of thatched cottages with its castle a dominant feature across the Wild Brooks of the Arun Valley: one of a number of 'spring-line villages' scattered like jewels on a necklace along the base of the chalk escarpment through West Sussex into Hampshire. And to the east, Southease, with one of the only three round towered churches in Sussex, all to be found in the Ouse valley; and Alfriston, one of England's most beautiful villages best known for the view of its 14th Century church across the water meadows of the Cuckmere. The idyllic environment of the South Downs settlements attracted scores of writers in the first half of the 20th century. In 1916 Virginia Woolf made her home in the eastern Downs at Rodmell and her sister's home at the Charleston Farmhouse became the summer gathering place for the Bloomsbury set.

There are four market towns. Travelling eastwards, the first is Petersfield, dominated by Butser Hill to the south, the highest point on the South Downs ridge at **889 ft. (271 metres)**, and the Hampshire Hangers to the east. It was founded in the 12th Century and gained its prosperity from its position on frequently travelled trade and pilgrimage routes, most notably those from London to Portsmouth and Winchester to Chichester, and from its place in the wool trade. A weekly cattle, sheep and horse market took place in its square until the 1950s. Most famous was the Taro Fair which took place every autumn and gradually shifted from a place to buy and sell stock to a popular funfair. It was located on Petersfield Heath with its lake and 4000 year old burial mounds: one of the most important Bronze-age lowland burial sites in the country. Although the stock market has gone, markets still take place in the town square, overlooked by the dominant statue to William III. The old drovers' routes now provide easy access into the surrounding countryside for walkers and cyclists.

Midhurst like Petersfield, lies on the River Rother. It is a fascinating small town known for its rich history. It possesses a fine array of ancient buildings. One of the oldest is the 15th century coaching inn, the Spread Eagle Hotel, which sits amongst attractive 16th century buildings in and around the old market square. The main shopping centre in North Street, through which the main highway passes, has somehow retained many of its 17th and 18th century buildings. Set in the splendid landscape of Cowdray Park, the Cowdray Heritage Ruins are all that remains of one of England's most important early Tudor houses after a major fire destroyed much of it in 1793. Located on the opposite side of North Street to the ruins is the Old Grammar School which was established in 1672. H.G.Wells was a pupil teacher there. Its buildings now house the South Downs National Park headquarters. The buttercup yellow window and door frames seen on a number of buildings around the town denote their association with the Cowdray Estate.

Some 6 miles (10 kms) to the east of Midhurst and to the north of the River Rother lies Petworth. This perfectly preserved market town is well known for its art, culture and many antique shops. It is dominated by the 17th Century Petworth House, inspired by the Baroque palaces of Europe. Its extensive grounds and deer park were designed by Capability Brown. The house, now owned by the National Trust, contains an impressive art collection. But it is the association with J.M.W.Turner that is perhaps most significant. The artist spent long periods at the mansion under the patronage of the 3rd Earl of Egremont and his paintings of Petworth Park can be seen there today.

Like Arundel and Steyning, Lewes is a 'gap' town that is located on a spur of downland elevated above a major river as it flows through the Downs. The town, the largest in any national park in the UK, is prominent in views from the Ouse valley and surrounding downland but is itself dwarfed by the majesty of the hills that surround it on three sides vividly described by William Morris as: *'lying like a box of toys under the great amphitheatre of chalk hills.'* A national gem of a market town with its knapped flint and many historic buildings, it is the county town of East Sussex and boasts a castle and formidable prison.

Its extensive conservation area encompasses the historic core of medieval and Georgian architecture, with striking views of the Downs from the High Street. It has a rich history: it is the site of the Battle of Lewes in 1264 when Simon de Montfort's forces prevailed over the forces of Henry III establishing the foundations of our modern Parliament. Thomas Paine, a founding father of the American Revolution, first propounded his revolutionary politics as a resident of Lewes in 1768. It is now known chiefly for the largest and most famous Bonfire Night spectacle in England, commemorating not just the Gunpowder Plot but the seventeen Protestant martyrs burnt at the stake there in the 16th Century.

Lewes is the hub of a rich variety of wildlife habitats from woodland to marshes, from floodplain to superb chalkland on the surrounding Downs. To the south-east, the abandoned chalk quarries at Southerham drew the attention of Dr Gideon Mantell, a local medical doctor and one of Britain's pioneering palaeontologists of the 19th Century. Here he collected a remarkable array of ammonites and fossilised sea urchins and fish which provided a greater understanding of the origins of the South Downs.

Like other market towns, Lewes once existed to support the agricultural economy of its surrounding area. Throughout history that agriculture has responded to the vagaries of the economic climate. The numbers of sheep and cattle varied dramatically as did the production of corn. Almost a century ago the chalk downland lay desolate and ungrazed as the economy crashed in the Great Depression. Immediately war was declared in 1939, thousands of acres were ploughed up and the corn produced made a major contribution towards feeding an isolated Britain faced with the prospect of invasion. As the tide turned, huge areas of the South Downs were requisitioned by the War Department for military training in preparation for the D-Day offensive. The land itself again became overgrown with scrub and coarse grasses. In 1946-47 the fields on Amberley Mount were still littered with spent mortar shells and was populated by hordes of rabbits that had honeycombed the hedge banks with their burrows. Kingsley Vale had to wait until 1990 for the clearance of the 6000 bombs that still remained there.

The downland served another, inspirational, purpose during the Second World War. One of the most iconic posters circulated by the government to raise the morale and fortitude of the British population was a painting of the South Downs by Frank Newbould entitled 'Your Britain fight for it now'. It perfectly illustrates this glorious countryside and the passion of those who have strived to protect and enhance a very special corner of England.

Lewes embedded in the South Downs
© Anne Katrin Purkiss South Downs National Park Authority

© Crown copyright. Imperial War Museum

CHAPTER TWO

THE PIONEERS 1923 -1966

The arrival of the motorcar and the burgeoning railways after the 1914-18 war brought an avalanche of newcomers to the South Downs. Then came the speculators and an unregulated rash of buildings. The worst was the creation of Peacehaven, originally a 'weekend colony' of huts and bungalows without roads or sanitation or any semblance of sensitivity to the surrounding countryside. The whole of the coastal chalk downland between Brighton and Eastbourne was threatened with urban developments and ambitious plans to create new seaside resorts. Even Beachy Head and the Seven Sisters came close to being lost to opportunists. Fortunately, dedicated individuals and wider campaigns swept to the aid of this threatened countryside.

It was these threats that led to the Society of Sussex Downsmen being founded in 1923 with the object of conserving the special qualities of the Sussex Downs. Plans were afoot to build a new town, another Peacehaven, on the Crowlink Estate at Birling Gap: a dip in the white cliffs of the Seven Sisters. The only option was to purchase the land and the Downsmen launched a public appeal. Using radical methods for their day, including landing an aircraft in front of the crowds coached in for the occasion, they raised money but, by the final day, they were still £5000 short, more than half the asking price. William Campbell, a resident of Eastbourne, who had previously donated £2000, lent the money and later forwent repayment thus ensuring the estate could be purchased outright in 1927. The Estate was later donated to the National Trust.

At much the same time, Eastbourne Borough Council became so concerned about the ravages of unconstrained development that it persuaded Parliament to agree to it compulsory purchasing 4000 acres of open downland. It then raised the necessary funds by extra tax revenues and by 1929 the purchase was complete: 'to ensure the free and open use of the Downs in perpetuity.'

It was the visionary Sir Herbert Carden who persuaded Brighton County Borough Council that the Downs to the north of the town should be conserved both to protect the town peoples' enjoyment of them and the town's water supply.

Often buying the land himself before selling it on to the Council at no profit, his efforts ensured that today some 12,800 acres remain in the ownership of Brighton & Hove City Council. Perhaps his most famous purchase was in 1929, when he bought 190 acres of the deep V, north facing valley known as Devil's Dyke, which had for long been a popular attraction for Brighton citizens, with an aerial ropeway to attract the more adventurous. Devil's Dyke was sold to the National Trust in 1995.

Plans in 1934 to build a racetrack near Portslade, sponsored by that same Brighton County Borough Council, provoked such outrage from other councils in the area that East Sussex County Council sponsored a Private Bill in the House of Lords to stop it and protect the Downs. The Bill failed on an assurance that the new Town and Country Planning Act would suffice to protect the countryside. Meanwhile s34 agreements were entered into to buy out development rights, ensuring permanent land change.

Whilst these local efforts successfully protected the eastern Downs, others were setting out to protect the countryside more generally. In 1926, a group of conservation-minded organisations formed a national body, the Council for the Protection of Rural England. The Sussex Downsmen became a paying member of the newly formed organisation and received much support in its own more localised efforts from this national body. The Council later became a charity in its own right, now known as CPRE, the Countryside Charity. I refer to it in this book as CPRE to distinguish it from its associated but independent county branches which include CPRE Hampshire and CPRE Sussex.

It was largely at the instigation of CPRE that Ramsay McDonald, the newly elected Prime Minister, established in 1929 a committee under the chairmanship of Dr Christopher Addison to consider the establishment of national parks. In his evidence to the committee on behalf of CPRE, Sir Patrick Abercrombie proposed twelve areas that might be most suitable for national park status suggesting as he did so that:

'Of the twelve areas given in the list, the High Peak and the South Downs would appear to have the first claim from the population point of view; from the point of view of national interest and intrinsic beauty the Lakes, Snowdonia, Exmoor and Dartmoor.'

The Addison Committee reported in 1930 recommending that national parks be created: *'to safeguard areas of exceptional natural interest against disorderly development and speculation'* and *'to improve means of access for pedestrians to areas of natural beauty.'* Unfortunately, the General Election in 1931 and the subsequent change of government, coupled with the ensuing economic crisis, put paid to the Committee's ground-breaking recommendations.

With the shelving of the Addison Report, the move to create national parks suffered a severe setback. But in 1936 the Standing Committee on National Parks was established: a group of individuals and voluntary organisations that championed the cause of protecting our finest countryside and ensuring access for all. John Dower played a key role in this committee and drafted much of its output. Born in Yorkshire in 1900, he was a keen walker and a great lover of the countryside. His great passion was always the fells and dales of his childhood. His work on the Standing Committee influenced the wartime Government's proposals for a better Britain. This culminated in the publication in May 1945 of a report that became known as the 'Dower Report'. The report proposed that a national park should be:

'An extensive area of beautiful and relatively wild country in which, for the nation's benefit and by appropriate national decision and action, (a) the characteristic landscape beauty is strictly preserved, (b) access and facilities for public open-air enjoyment are amply provided, (c) wild-life and buildings and places of architectural and historical interest are suitably protected, while (d) established farming use is effectively maintained.'

John Dower wrote:

'I should have included at least two southern areas in Division A (suggested national parks) or B (reserves for possible future national parks) if I were not reasonably satisfied that they would, in future, be adequately dealt with by other agencies, the South Downs by the county and local authorities....'

Consequently, he classified the South Downs as an 'other amenity area.'

In July 1945 Sir Arthur Hobhouse was asked to chair the National Parks Committee and to take forward the recommendations in the Dower Report. He had been a member, and latterly chaired, Somerset County Council and was chairman and president of the County Councils Association. The terms of reference for his committee, whose membership included John Dower, incorporated the selection of areas to become national parks and the necessary administrative arrangements. In September 1945 Hobhouse had his first meeting on the subject with the responsible Minister in the Ministry of Town and Country Planning. Hobhouse specifically asked if, in determining future national parks, his committee should confine itself to the recommendations in the Dower Report. Tellingly, so far as the South Downs was concerned, he was told no, and that he should take a pragmatic approach to how many national parks he should recommend.

In July 1946 six members of the National Parks Committee, including Sir Arthur Hobhouse Lord Chorley and Dr J S Huxley, undertook a two-day survey of the South Downs. On the first day their journey covered Winchester, Petersfield, Harting Down, Chichester, Arundel Park, Steyning, Bury, Bignor, Goodwood Park, Treyford, Liss, Alton and Selborne. On the second day they visited East Meon and West Meon, Butser Hill, Buriton, South Harting, Harting Down, Goodwood Park and Chichester, Petersfield and Winchester.

In their report they stated that:

'The South Downs appear to be in a different category (to the Berkshire Downs) owing to the steeper slopes of its escarpments, the large areas under mature hardwoods, and the greater variety of distant views due to proximity to the sea. With the exception of Beachy Head and Ditchling Beacon, which are the special concern of Eastbourne and Brighton and are protected as such, the South Downs lie mainly within the area of West Sussex. They are at present protected as a landscape zone under the West Sussex County Council Planning Scheme and by arrangement with the owners, who happen to be relatively few and large. The area is at present free from disfiguring development and its woodlands have so far escaped, in the main, felling and replanting.'

'The extent of use by the public appears to be small and principally confined to motorists, using the main thoroughfares to the sea. This area is, however nearer to London than any other considered for the purpose of a national park and it might presumably be developed for that purpose. No final decision was reached by members who viewed the South Downs, but the general opinion was in favour of this area if a national park is to be created within easier distance of London.'

'The National Parks Committee considered Professor Tansley's proposals to include a large area between Petersfield and Pulborough extending northwards almost to Farnham and Godalming. It was felt that, although there were pockets of characteristic Sussex countryside within Professor Tansley's extension, the quality was not sufficiently consistent, nor was the character suitable, to justify its inclusion in a downland national park.'

'The precise boundary of the South Downs was not decided but it was thought that the Eastern end should be excluded up to the valley of the Adur and left for the protection of Brighton and Eastbourne, and that the western end should be extended to the neighbourhood of Winchester and be extended northwards to include Selborne.'

In 1947 the Hobhouse Committee recommended that the Government should create twelve national parks in England and Wales, including one for the

South Downs with a boundary incorporating an area from around Selborne in the West to Eastbourne:

'We were impressed with the importance of including at least one national park within easy reach of London. There exists in the South Downs an area of still unspoilt country, certainly of less wildness and grandeur than the more rugged parks of the north and west but possessing great natural beauty and much open rambling land, extending south-eastward to the magnificent chalk cliffs of Beachy Head and the Seven Sisters. We recommend it unhesitatingly on its intrinsic merits as well as on the grounds of its accessibility.'

Based on the report of the Hobhouse Committee, legislation was brought forward which paved the way for The National Parks and Access to the Countryside Act that was enacted in 1949. Britain's first ever national park, the Peak District, was designated in 1951. The South Downs was meant to be in a second tranche of designations. But in the meantime, a government drive to increase agricultural production led to large areas of precious chalk grassland being cultivated.

The National Parks Commission was established through the 1949 Act to co-ordinate government activity in relation to national parks. In 1957 the Commission announced that:

'The South Downs, recommended for designation as a national park by the Hobhouse Committee, has been engaging our attention for some time and the area was visited by survey parties from the Commission in May and July 1956. The recreational value of the South Downs as a potential national park has been considerably reduced by extensive cultivation of the Downland, and, after our inspections, we concluded that the designation of this stretch of countryside as a park would not be appropriate. At the same time the region has great natural beauty, and its ready accessibility from London makes it especially vulnerable to development. We are accordingly proposing to consider the designation of the area as one of outstanding natural beauty and we hope during the coming year to open discussions with the local authorities concerned.'

In his 1945 Report to Government on National Parks, John Dower had also suggested that there was need for the protection of certain naturally beautiful landscapes that were unsuitable as national parks owing to their small size or lack of wildness. His recommendation for the designation of these 'other amenity areas' as Areas of Outstanding Natural Beauty (AONB) was eventually embodied in The National Parks and Access to the Countryside Act of 1949.

AONBs were to be designated on the same basis as national parks in terms of their landscape value. However, they did not have to meet a second criteria necessary for national park designation, that of providing recreational opportunity. Work began on designating them in the 1950s and, like national parks, this was split into tranches.

Dr Francis Rose was Britain's finest field botanist of his generation, arguably of the last century. He had been engaged in the 1940s in identifying areas for special protection and was involved in the designation processes for the Sussex Downs AONB and East Hampshire AONB. He recalled that he did not agree with the Hobhouse Committee's decision to exclude from the proposed South Downs National Park the area to the north and east of the chalk to the Surrey border, the Western Weald, believing the area was deserving of inclusion not only on its own merit but also because of its connectivity to the chalk hills. This view echoed that within the Hobhouse Committee by Professor Tansley but rejected.

Having been rejected for national park status, the process began for designating a Sussex Downs AONB. When it came to consideration of the boundary, this time the rationale for including the Western Weald was accepted so, unlike the proposed national park, the AONB was to encompass not just the chalk hills in East and West Sussex but also the sandstone ridges and clay vales of the Western Weald in West Sussex. Although the Sussex Downs AONB was in the first tranche for designation this was not achieved until 1966, some nine years after the process began, the delay being caused by 'administrative issues.'

The East Hampshire AONB was put into the second tranche and work began on its designation in 1958. The visiting party from the National Parks Committee recommended that a much greater area than the proposed South Downs National Park was worthy of designation, stretching nearly to Winchester. Their recommendations were accepted by the Committee and then discussed with Hampshire County Council's Chief Planning Officer. He agreed largely with what was proposed but suggested the town of Petersfield and the village of Liss should also be included and that this would make for a neater boundary with the co-terminus Sussex Downs AONB. Given the size of Petersfield, the informal views of the Ministry of Housing and Local Government were sought. Although the response was not encouraging, the Committee pressed ahead with this inclusion and others. The East Hampshire AONB was designated in 1962.

Despite having been considered for national park status since the late 1920s, and with many twists and turns on the way, finally both the chalk hills and Western Weald were formally acknowledged as nationally important landscapes, worthy of protected status, albeit as AONBs and not a national park.

CHAPTER THREE

AN IMPERFECT HAND

Though national parks and AONBs were designated under the same legislation, and shared a common purpose of conserving and enhancing the landscape, wildlife and cultural heritage of the areas concerned, from the start their funding and administrative arrangements differed to the detriment of AONBs.

The Hobhouse Committee envisaged a central body, the National Parks Commission, which would set policy in relation to national parks. The Committee recognised that planning could not be undertaken by a central body: it needed to be based locally. However, to achieve the Committee's ambitions, each national park needed to be treated as an entity for planning purposes and not split by administrative boundaries. They therefore proposed a solution which achieved this result whilst conforming to the planning system proposed for the country as a whole brought in by the Town and Country Planning Act 1947. Thus, the Commission would have a part to play in establishing national park committees and would appoint half the members of the committees and the chair. The remaining members would be appointed by the relevant county councils.

This proposal was watered down in the 1949 National Parks and Access to the Countryside Act to the chagrin of those who wholeheartedly supported the concept of national parks. Thus, the National Parks Commission was required to consult with all relevant local authorities about the arrangements for planning. The membership of the board was to be two-thirds from county councils and one third appointed by the Commission.

This arrangement satisfied hardly anyone. In 1956, Sir Arthur Hobhouse and several of his colleagues from his Committee signed a letter to The Times complaining that only in the Peak District had a properly functioning joint committee been formed. The arrangements did not suit county councils either, who complained to the Churchill Government of their potential loss of power. And, in the 1970s, there were complaints from district councils that they were excluded from the arrangements, despite their role in planning outside the national parks.

By the time of Local Government reform in 1974, only the Lake District and Peak District were managed by independent boards who employed their own staff, could own land, and raise monies from county councils.

The work of the National Parks Commission was then absorbed by a new body, the Countryside Commission. Following an agreement reached around that time between the Association of County Councils and the Countryside Commission, the remaining five national parks in England were each managed by a committee of their respective county councils, to which the determination of planning applications, and 'countryside functions' were delegated. Planning policy varied from park to park. Thus, most of the national park boards were effectively controlled by their county councils.

Though these administrative arrangements were complex and differed from national park to national park, the financial ones were comparatively simple. Three-quarters of the annual expenditure was provided by a direct grant from central government, with the remainder being channelled through the relevant local authorities.

All these arrangements were far from satisfactory, but national parks fared far better than AONBs. Despite much encouragement from the Countryside Commission during the 1970s and 1980s for county councils to set up joint advisory committees where several local authorities were involved, nothing was forthcoming in the way of financial support to promote their set up. Although planning policies recognised the existence of AONBs, these were the responsibility of the relevant local authority and differed from district to district and borough to borough.

With no formal process or outside funding available for AONBs, it was up to the county councils to make their own arrangements. In the Sussex Downs it was East Sussex County Council who led the way. In 1981 it set up a project to manage the Heritage Coast, the South Downs Conservation Project, which eventually covered the whole of the area of the AONB within its jurisdiction. Two full time officers managed this project. It would have been impossible to find two greater enthusiasts, or larger than life characters, than Paul Millmore and Phil Belden. Paul was a gritty Yorkshireman whose family had owned a wool mill in Bradford. Bearded and resplendent in his trademark red braces and bow tie, he was a forceful and distinctive presence in the story of the South Downs until his death in 2012. Phil, also bearded, a tall man with unabated enthusiasm for the South Downs and the countryside more generally, was equally dedicated to the job in hand. They threw themselves into the work and, to increase their effectiveness, Paul formed a volunteer ranger service, modelled on those to be found in national parks elsewhere, to help with countryside management and to improve the rights of way in the area.

From time to time, the Countryside Commission provided funding to support some of their projects and became sufficiently impressed with what was being achieved to encourage West Sussex County Council to set up a similar project there. A Statement of Intent was signed by the two councils in 1986 and a Sussex Downs AONB forum was launched to discuss matters of interest regarding the AONB. Phil Belden moved from East Sussex in 1988 to spearhead activity in West Sussex and he set up a voluntary ranger service there in 1989.

As well as the occasional financial support from the Countryside Commission, both county councils were fortunate to be able to take advantage of the Manpower Services Commission scheme set up to help tackle the unemployment crisis in the 1980s. Although this was a short-term arrangement it was highly productive whilst it lasted. In East Sussex two fully mobile project teams were able to service the rights of way and undertake conservation projects. Their work was supplemented by technical support teams with additional teams called upon for specific tasks. In West Sussex similar teams tackled major projects such as clearing the scrub and erecting fencing to enable the Devil's Dyke to be grazed by animals once more.

Whilst the Sussex based county councils made some progress, agricultural practices were out of their hands. The extensive ploughing of land, referred to by the National Parks Commission in the 1950s, had continued. Swathes of ancient chalk grassland were converted to arable and intensively farmed. The fertilization of the remaining grassland, whilst providing more food for the sheep and cattle, led to the more delicate and slow-growing plants being crowded out. Insects such as the beautiful Adonis Blue and Chalkhill Blue butterflies that were dependent upon fine chalk grasslands habitats, disappeared from their prime sites. The special qualities of the downland that had evolved through 5000 years of mostly benign management were almost obliterated.

The writer and campaigner Marion Shoard's book 'The Theft of the Countryside', published in 1980, set out the consequences of the expansion and intensification of modern agriculture and this sparked a public debate on the issue.

To illustrate the lack of any real public say over the fate of valued landscapes, she highlighted the case of Graffham Down, a stretch of downland above the village of Graffham in West Sussex which had been unimproved chalk grassland rich in biodiversity. A new owner, intent on maximising food production, cleared

On Mount Caburn, one of the few rich chalk grasslands that evaded the intensive farming on the South Downs
© *Malcolm Emery Natural England*

away most of the valued features and turned it into a barley prairie. She wrote:

'Until recently one of the gems of the Sussex landscape, its unhappy fate demonstrates as typically as anything the impotence of the community in the face of agricultural change.... The butterflies, the flowers, the wild animals, the birds, the seclusion of the woods and the peace of the downland made Graffham Down an unusually splendid place to ramble, picnic or play. But to the farming community Graffham Down merely constituted uncultivated marginal land, unused capacity for food production.'

Marion Shoard contended that if such activity had been under planning control this could have been avoided. Less controversially, or so it seemed at the time, she contended that even without such controls there would have been a better chance to save Graffham Down had it been within a South Downs National Park with a better resourced authority to resist the site's transformation.

In response to the public outrage at the destructive farming practices in the UK, a new Agriculture Act in 1986, with funding from the European

Union, made it possible to establish Environmentally Sensitive Areas (ESAs). Landowners were to be paid to enhance landscapes and wildlife. Thanks to the East Sussex Conservation Project and its dedicated officers, and a good deal of lobbying, the Eastern Sussex Downs ESA was one of only five ESA schemes launched nationwide in 1986. To benefit from the scheme farmers were required to maintain traditional farming practices and to restore downland grassland habitats. West Sussex County Council then lobbied for the ESA scheme to be extended to West Sussex and into Hampshire. However, these projects were only partially successful. Though they led to the conversion of arable to grassland, that grassland was by no means the species-rich chalk grassland of the past but often a boring monoculture of rye grass with clover. It took further lobbying over several years to change the ESA scheme to ensure a better outcome.

The South Downs faced other threats. The Brighton Bypass cut through swathes of chalk downland and an even more destructive bypass around Worthing was mooted. At Winchester, just beyond the boundary of the East Hampshire AONB, the proposal for the M3 cutting at Twyford Down was coming close to reality despite fierce opposition.

In March 1989 the Times published a controversial article by Marion Shoard that made the case for six new national parks, including one for the South Downs. She pointed out that the AONB designation was *'a designation that brings few benefits and is now perhaps best abandoned.'*

Presaging what was to come, the West Sussex County Council's Planning Officer responded by questioning whether a national park was *'right for this soft and sensitive landscape'* and if this *'status posed a threat to the people who manage the land so well.'*

CHAPTER FOUR

THE FORMATIVE YEARS
1990-1992

From my perspective, the South Downs Campaign can trace its genesis to an announcement by the Conservative Environment Minister, Chris Patten, on 13 December 1989. It was timed to coincide with the 40th anniversary of The National Parks and Access to the Countryside Act. He said that a National Parks Review Panel was to be set up under the chairmanship of Professor Ron Edwards. This became known as the Edwards Review, and I will use that title for it from now on in. This announcement created a vigorous public debate in the South Downs as elsewhere. Consequently, a conference was convened by the Sussex Rural Community Council and CPRE Sussex on 9 February 1990 to consider the question of a national park in the South Downs.

The conference was opened by Amanda Nobbs, the Director of the then Council for National Parks. She offered a national perspective on the case for the South Downs to become a national park and how this could enrich the national park family. She suggested that the reasons against designation in the 1950s had been eclipsed by the benefits national park status could offer 40 years later and how the South Downs could be a pioneering role model to promote better environmental practices. She went on to suggest that national parks could enjoy greater appreciation if they were not perceived as remote and marginal, as they were by the largely south-east based media, and how bespoke legislation passing through Parliament to give the Norfolk and Suffolk Broads equivalent status to a national park was already breaking the mould.

I well remember the speech given to the conference by Marion Shoard. It seemed to me that she set out with great clarity the rationale for why the South Downs should receive the highest level of protection and the benefits that could accrue from that designation. Having re-iterated the views expressed in the Hobhouse Committee about the area, she moved on to criticise the devotees of upland country who had dominated the countryside movement in the middle of the 20th Century, John Dower being particularly influential. Although Parliament had rejected Dower's proposed definition of countryside suitable for national park designation through the National Parks and Access to the Countryside Act 1949, replacing it with one wide enough to embrace lowland as well as upland areas, it was Dower's thinking which tended to hold sway in the National Parks Commission. As an example she cited the case of the widow of John Dower, Pauline Dower, the longest serving member of the Commission, who had told the Town and Country Planning Association in 1952 that *'The first duty of the National Parks Commission is to select those areas of England and Wales which would, in their opinion, fulfil the accepted definition given by John Dower in his report'* despite Parliament, through the 1949 Act, requiring the Commission to apply different criteria.

Marion Shoard pointed out that whilst mountain and moorland can be found all over the world, chalk downland is largely an English speciality, unknown elsewhere apart from odd outcrops in France, New Zealand and a few other places. Furthermore, she proclaimed that the South Downs, with their bold, clear line and dramatic northern profile, are the finest of the several chalk ranges radiating out from Salisbury Plain. Compared to lowland landscapes like the South Downs she said: *'Our ten national parks solemnly enshrine landscapes which are not only less threatened but which are in global terms less important.'* The Downs met the criteria for national park selection: natural beauty and the opportunities they afford for open-air recreation, having regard both to their character and to their position in relation to centres of population.

Marion argued cogently that the better funding and administrative arrangements that national park designation would bring could make a huge difference to the area. She cited the Countryside Commission's recent announcement that it would pay only half the cost of a single person for each AONB until 1992, against the longer-term commitment to national park funding covering a team of officers to plan and manage the area. She concluded by suggesting that *'It would be easiest to create a national park out of the area covered by the Sussex Downs AONB. This area which covers 379 square miles, including a small stretch of Wealden scenery, north of the escarpment in West Sussex and lying between Petersfield, Petworth and Haslemere.'*

I was persuaded by her vision.

At that time, I chaired the Sussex Wildlife Trust and recall that at the beginning of May 1990 Penny Edwards, then the Trust's Director, discussed with me a proposal that there should be a meeting regarding a South Downs National Park at our headquarters at Woods Mill, situated to the east of the Adur valley. This initiative came from Phil Belden, who had by this time become a trustee of the Sussex Wildlife Trust as well as continuing his role with West Sussex County Council. Penny circulated a letter which stated that the object of the meeting would be to decide how the designation of the South Downs as a national park could best be promoted and what the composition and structure of a group, with such a role, should be. She also suggested that I, as chairman of the Trust and as host, should chair the meeting which she would also attend.

A small group of seven duly met on 15 May 1990. Brian Short from the Sussex Archaeological Society and University of Sussex and David Harvey from the Nature Conservancy were present. Phil Belden and Paul Millmore were attending in an unofficial capacity. Phil and Paul had rapidly concluded from their experience with East and West Sussex County Councils that the only certain way of ensuring the South Downs were properly conserved and enhanced was for the area to become a national park and had devoted much time and energy to this end. Their efforts were astounding though, I have to confess, unknown to me at the time. They had not only attempted to convince their respective councils but had also given many talks in their own time and engaged with numerous environmental organisations. They had raised the profile of the issue at national level through interviews and letters to the press and had engaged with the Council for National Parks. It was Paul and Phil who suggested that Amanda Nobbs should also be invited, something to which I readily agreed. Apparently, they had approached her immediately after her speech to the February conference, offered her a South Downs National Park pin badge, and described all their impressive local efforts to set up a campaign group. Amanda recalls shaking hands on it and agreeing: *'We are going to do this!'* little appreciating the tortuous route ahead.

The Council for National Parks was to be crucial in the years ahead, and Amanda Nobbs an indomitable and inspiring presence for the next 10 years, so it is worth introducing them here. It is also worth relating the background to the Chris Patten announcement about which I was wholly ignorant at the time and for many years to come. I am indebted to Amanda for providing this important part of the story.

The Council for National Parks had its origins in the 1936 Standing Committee on National Parks and had evolved into a charity promoting more effective management and better protection for our finest landscapes in England and Wales. In 1977 the Committee became the Council for National Parks, which later morphed into the Campaign for National Parks. From here on I will use the initials CNP to avoid confusion. By the 1980s, it had representatives from about 50 environmental and amenity organisations and an observer from each of the national park authorities.

Amanda Nobbs had joined the organisation in 1987. Whilst studying for a Masters in Conservation, she had been taken by the opportunity to recover the wildlife richness of vulnerable downland landscapes. She records: *'When asked over a dinner attended by national park cognoscenti to reflect on my impressions after 12 weeks in post, I suggested the time was right to look again at the potential for lowland national parks, and in particular the South Downs.'*

The response was not favourable. The idea that a downland landscape could offer the qualities associated with a national park was met with resistance. Some thought the Downs insufficiently rugged. Others thought the priority should be better measures in areas already designated as national parks.

The idea was given a mixed reception too at CNP, at least initially. The strongest support came from a number of those who had been involved in the post war campaign for national parks. They appreciated that the South Downs had the necessary landscape quality and saw ample opportunity for recreation. Amanda told me that she had taken heart from a hand-written note from Lord Hunt of Everest fame, a past president, who was convinced this was a cause worth pursuing. Lord Hunt wrote movingly of his experience of looking out across the Downs when war was declared with Germany, drawing inspiration and comfort from the majestic rolling landscape amid the fear that nothing would be the same again. If a person famed for his ascent of Everest thought the South Downs had the qualities of a national park that was good enough for her.

Subsequently, CNP resolved to pursue a twin track approach of calling for more effective arrangements for park authorities and for new national parks. At this stage the candidate areas the Council had in mind were the North Pennines, the Cambrian Mountains, the New Forest and the South Downs.

An event at Chatsworth House in the Peak District on 20 September 1987, provided an opportunity to promote these ideas more widely. The event marked the culmination of a government campaign to raise the profile of national parks. Brian Redhead, the Radio 4 Today programme presenter and president of CNP, was invited to give a keynote speech. Brian described national parks as *'Not ours, but ours to look after'* and ended with a call for *'more money, more powers and more parks'*. The then Secretary of State for the Environment, Nicholas Ridley, was unreceptive, but his successor Chris Patten picked up the message and duly made his far-reaching announcement in 1989 about the setting up of the Edwards Review.

At our first meeting in 1990 we felt that the Government might be looking for quick, tangible ways to demonstrate action on the environmental front and were encouraged that Chris Patten had indicated in his announcement that new national parks would be sought. We decided our best approach to getting a national park might be through the framework of the 1949 Act, which seemed to allow enough flexibility for tailor-made solutions for the specific problems of the South Downs. We realised there were powerful voices that did not share our vision: neither the local authorities nor the Society of Sussex Downsmen were in favour and the farmers seemed unimpressed by the merits of the case.

We decided to call our group the 'South Downs National Park Campaign' and I was invited to chair it 'for the time being'. At that first meeting we laid down an approach which was to be pivotal to our success; the melding of national and local endeavour. National organisations had access to Ministers, to Whitehall and to opinion formers that locally we lacked. But a campaign of the sort we mounted, that endured over so many years and overcame so many setbacks, could only have succeeded if local people shared the vision and were prepared to support it. Most of the subsequent legwork was undertaken by local volunteers, giving up their time because they cared so much for the area. That is not to suggest that those representing national organisations were not fully engaged too. Amanda Nobbs told me that she reflected on her way home from that first meeting that a national park was not an idea that could be imposed. So, as well as making appearances on Country File and Farming Today, she joined me and others in talking to local groups, in sometimes chilly village halls, to help spread the message.

By the time of our second meeting in June 1990 the Edwards Review had launched a consultation which, we were disappointed to note, had not engaged with many small and local environmental or amenity groups. At that meeting Peter Brandon joined us representing The Society of Sussex Downsmen. He stated that the Downsmen were *'unable to campaign on the national park issue because their initial response to the consultation had been against designation. They are adopting a low profile on the issue and are therefore declining the invitation to join the Campaign.'* It was difficult for me to understand why the Society of Sussex Downsmen were not whole-heartedly in favour of a national park, given their distinguished record in protecting the South Downs in the 1920s and 1930s. Peter Brandon, probably the greatest authority on the history and geography of the Downs and a great lover of the Sussex countryside, remained on our committee in his own right from 1990 until 1995.

In July, we welcomed Richard Reed, who had been an active member of the Sussex Downsmen since 1947 and later became its chairman. Acting as an observer, he was always very friendly and constructive despite the stance taken by the Downsmen. His stunning photography of the Downs was of great value to our Campaign in the years to come. We also welcomed Andrew Lyall from Cuckfield, who came as a representative of the Ramblers Association. Andrew was an extremely able retired civil servant who informally acted as our honorary secretary after Penny Edwards left the Sussex Wildlife Trust. He was hugely supportive. We were very fortunate in having him and he made a significant contribution to our work over several years. Ian Elliott, a member of West Sussex County Council, also joined us at this meeting. Initially he appeared to be supportive, but when he became chairman of that council, he left our committee and became vociferously opposed to a national park.

In these early days we had to address the question of how our organisation should be structured. First and foremost, we took the decision that we would be a campaign of like-minded groups. Given our lack of resources, we simply could not have coped with individual membership, even if that had been what we wanted. Although we operated for some time as a small committee, we nevertheless drew up a formal constitution. The appointment of a campaign officer was also considered, but we were struggling to obtain funding. Sussex Wildlife Trust were generously paying half of the costs, which were mostly spent in producing and mailing literature. A significant sum provisionally offered by CPRE was not forthcoming. Our attempts to extract £5000 from East Sussex County Council, which had been raised by the voluntary rangers, also failed. So, for the first few years of our campaign, the Sussex Wildlife Trust held our money and paid our bills. When their auditor eventually decided that the Trust should not manage funds over which it had no control, we opened our own account. I found myself by accident taking on the added task of honorary treasurer. It was a modest commitment at the beginning, but it grew once we needed to raise significant funds for the employment of an officer.

Perhaps surprisingly, we were swayed by siren voices, into abandoning all-out unconditional support for a national park to instead supporting a set of ideals which could be promoted within a framework of a national park structure. We therefore changed the name from 'South Downs National Park Campaign' to 'South Downs Campaign Group'. We hoped this nuanced change of words would entice the likes of the Society of Sussex Downsmen to join our campaign. We were mistaken. Later we changed the name again to 'the South Downs

Campaign'. To save confusion I shall refer to it as 'the Campaign'.

We drafted a mission statement proposing a better deal for the South Downs. In it we called for a more unified administrative body for managing the area that would have strengthened planning powers. We pressed for greater funding for environmentally friendly farming and for maintaining the existing fabric of the Downs, especially landscape and archaeological features. We also identified the need for greater resources for maintaining the villages as vibrant living communities. Finally, we pressed for better recreational management.

One of the main arguments against a national park, voiced loudly at that time, was the fear that the creation of a national park would increase the popularity of the South Downs for recreation to such an extent that it would negate the very purposes of designation. This was an important local issue which we acknowledged in the early debates within our campaign committee.

At this time, we were entirely focused on the chalk hills in Sussex, though we appreciated that they extended well into Hampshire. The reasons for this were fourfold. First, this was essentially the area proposed for national park status by Hobhouse. It is the case that not all national parks have boundaries that follow his review's recommendations: Northumberland, most notably, significantly differs from what was proposed. But at this time, we felt the Hobhouse boundary should be our starting point. Secondly, we were influenced by several notable figures: David Streeter, an academic who sat on the Edwards Review, Adrian Phillips, a member of the Countryside Commission and Fiona Reynolds, Director of CPRE and later of the National Trust, apparently a fixture on reviews of national parks. All appeared to favour a concept based on the American model of remote and seemingly wild places. We felt that the chalk hills best reflected this concept. Our third reason was a pragmatic one in that our group at the time was made up of people from Sussex and we were not aware of a similar push in Hampshire for a national park.

But our fourth reason was the most pressing at the time. As I wrote in my introduction to the Sussex Wildlife Trust's 'A Vision for the South Downs' which the Trust produced as part of our Campaign:

'History reveals that the comforting elegance of the chalk landscape of the Sussex Downs has been continuously modified in cycle with the fluctuation of Mans' need, in time of war, peace, boom and depression.

The integrity and natural beauty of the Downs withstood all the vicissitudes of history up to 1945. In the last fifty years the past has come close to being obliterated by modern farming practices driven by runaway agricultural policies. Urban development has added to the destruction. It is only through good fortune, the natural defences of the steepest hillsides, which withstood the onslaughts of the modern tractor, and the diligence of some individuals, that there is anything left of the very special grasslands and the ancient woods.'

It seemed clear to us that AONB status had been inadequate for conserving and enhancing the species-rich chalk turf and so that was where we focused our attention.

In September 1990 we wrote to the Countryside Commission about our aspirations for the South Downs. We told them we were seeking the establishment of *'a representative and local statutory body of equivalent status to a national park authority'* for the South Downs between Eastbourne and Winchester. Knowing that the views of the farming community would be important, we reassured the Commission that we had met with their representatives: the Country Landowners Association, the National Farmers Union and the Farmers and Wildlife Advisory Group. We had been delighted to find that there was considerable common ground between us. However, a fundamental stumbling block in many farmers' minds seemed to be simply the name. Farmers feared that a national park would attract masses of new visitors and that would damage their interests and be unacceptable to them. Although we could not subscribe to that view, we considered that a satisfactory solution would be to drop the term 'National Park' and simply use 'Authority' as had been done in the Broads. We appreciated that this could only happen through an amendment to the 1949 legislation or through creating special legislation as had happened with the Broads, though for very different reasons.

Meanwhile, a Joint Meeting of the Environment and Coast and Countryside Committees of East Sussex County Council and West Sussex County Council and other local authorities, was reviewing the future management of the Sussex Downs AONB. A paper initiated by West Sussex County Council set out the options of a national park and a conservation board. It covered all eventualities by proposing that an AONB conservation board should be established, with the possibility of a review of the national park option after five years.

A conservation board for an AONB was, at that time, something of a novel concept and moved forward significantly the governance arrangements from what they had been. The proposal was that the two county councils would delegate their countryside management responsibilities to the new body. It was hoped that all the district councils would consult it on all planning applications outside towns and villages, with a right for the board to be heard at the relevant planning committee if there was disagreement, thus safeguarding the role of district councils in the planning process. The paper suggested that the Countryside Commission might be attracted to the model as a way of improving the management of AONBs more generally. Given the strong opposition to a national park being voiced within West Sussex County Council, it was not unreasonable to surmise that the proposal for a conservation board was, for some, also a means of seeing off this possibility. For ourselves, the conservation board proposal fell well short of what we thought was necessary for the South Downs.

We responded in September of that year with a paper entitled 'A Draft Outline for a New Style South Downs National Park Authority.' This six-page document included a map with a proposed boundary that covered the chalk Downs in Sussex and Hampshire. In the accompanying letter we said that it would be unwise to pre-empt the deliberations of the Edwards Review which was due to publish its report early in 1991.

A month later the Campaign hit the national press in a half-page spread in The Guardian which set out the pros and cons of a national park against a conservation board. The article stated that *'virtually everybody was in agreement that the Downs needed drastic action to save them from over-intensive farming, and the ever-increasing pressure from tourism and development.'*

This article appeared just before a crucial meeting of the Sussex Downs AONB Forum at which the results of a public consultation on the future of the South Downs was for discussion. To our annoyance the paper circulated ahead of the meeting failed to mention the Campaign, though it did reflect the views of our member organisations. More importantly, ahead of the meeting, the Regional Officer of the Countryside Commission wrote to West Sussex County Council setting out a summary position statement from the Commission in which it supported the creation of a conservation board for a period of five or six years.

Thus, at the Forum, which was supposed to debate the future status of the South Downs, *a fait accompli* was delivered by the Countryside Commission. We had prepared a two-page statement to be delivered at the meeting on behalf of the Campaign in which we declared that the Sussex Wildlife Trust, together with over thirty other conservation and amenity bodies in Sussex, as well as CNP, were convinced that the establishment of a tailor-made national park provided by far the best solution. My reaction on hearing of the Countryside Commission's statement was that the immediate future of the South Downs had been sealed and a statement by me at that stage would have been counter-productive. I therefore did not speak.

There was a lot of unhappiness among our member groups about the way the Forum had been conducted. Our committee decided I should write to the Countryside Commission, members of the Edwards Review, the Department of the Environment and to our own supporters. As my letter to the chairman of the Commission went unanswered we agreed that our next opportunity for major campaigning would be in response to the Edwards Review report.

Despite the stance taken by the Countryside Commission we garnered hope from the attitude of the main political parties:

The Conservative Government's White Paper on the environment not only looked forward to the Edwards Review report but stated:

'The Government will also consider any case for designating new national parks in suitable areas where landscape conservation and recreational opportunities can be combined, either under the 1949 Act, or by creating further tailor-made bodies like the Broads Authority.'

The Labour Party's programme for the environment 'An Earthly Chance' specifically mentioned the South Downs:

'We shall establish new national parks. Our first aim shall be to remedy the regional imbalance...by establishing new national parks in lowland Southern England. Early candidates will include the New Forest and the South Downs.'

And the Liberal Democrats agenda for the environment also supported the creation of new national parks.

At the same time, we understood that the internal discussions between the local authorities over the creation of a conservation board were not going well, with arguments over funding and planning controls. The question of a financial contribution from boroughs and districts for countryside management had not yet been addressed. None had indicated they were prepared to put money in. Chichester District Council, which had the largest portion of the Sussex Downs AONB within its boundaries, was reluctant to hand over any control to another body. Only the boroughs of Eastbourne, Brighton and Hove were willing to delegate all their planning powers in the limited area of AONB for which they were responsible. As regards development control, the other local planning authorities offered to 'consult' the conservation board on relevant planning applications 'outside settlement areas.' It was uncertain as to whether members nominated by the Countryside Commission would have more than advisory status. We considered that the proposals being put to the Countryside Commission were extremely weak.

On 21 March 1991, the eagerly anticipated Edwards Review report 'Fit for the Future' was finally published. It was a very thorough review that considered the role of national parks and their management since their inception. The report recognised that national park committees had faced serious impediments in pursuing their aims, and that national parks had not lived up to the expectations of their founders. The Panel therefore proposed a radical change to the status quo, that all national parks should be managed in future by independent national park authorities (NPAs) that would be unconstrained by the sometimes conflicting responsibilities that was currently the case.

Another very welcome recommendation in the report was that the role of national parks should be broadened and strengthened through a new 'National Parks Act' with three defined purposes:

1 *To protect, maintain and enhance the scenic beauty, natural systems and landforms, and the wildlife and cultural heritage of the area.*
2 *To promote quiet enjoyment and understanding of the area, insofar as it is not in conflict with the primary purpose of conservation.*
3 *In pursuance of these purposes, the national park authorities should support appropriate agencies in fostering the social and economic well-being of their communities within the national park, in ways which are compatible with the purposes for which the national parks are designated.*

The more important and far reaching of the report's recommendations would require primary legislation. We needed to play our part in pressing the Government to bring forward a new national parks bill. But in the meantime, we were much encouraged by what we read: the proposals were significant in bolstering the governance of national parks in a way that reflected our own views. If implemented, they would ensure that a new South Downs National Park could be established with an effective and independent board to oversee it. The proposals were certainly significantly different to what was currently under debate for a conservation board and again starkly demonstrated the gulf between the status of national parks and AONBs.

There was one element in the report that would come back to haunt us when we eventually reached the stage of having a Public Inquiry on the proposed South Downs National Park. That was their recommendation that new national parks must have '*an extensive tract of countryside which provides a sense of wildness.*' The report was also unhelpful to our cause in stating that '*Earlier reports (notably Hobhouse) have stressed the importance of having national parks accessible to main population centres, with the implication that southern Britain is poorly served. Since the reports were written, the personal mobility of the population as a whole has increased greatly, putting a much wider range of holiday and recreation areas within easy reach of urban populations.'*

By this time CNP had been joined by CPRE and The Ramblers Association as members of the Campaign, and the Sussex Wildlife Trust by the Sussex Archaeological Society and the Sussex Rural Community Council. A further 39 other organisations supported our aims. In a press release to be timed with the publication of the Edwards Review report we announced that we were writing a letter to the Parliamentary Under-Secretary for the Environment, calling on the Government to take immediate steps to establish a new-style national park on the South Downs, tailor-made to suit local needs. We said that we appreciated that the Edwards Review was unable to make specific recommendations on new national parks, other than the New Forest, because of time constraints. We welcomed their recommendation that the Countryside Commission should make an assessment of possible new national park areas in England.

Despite this public stance, in private we were extremely disappointed that the Review Panel had not supported a South Downs as well as a New Forest National Park. Amanda Nobbs advised us that Adrian Phillips, now Director General of the Countryside Commission, had given her the clear impression that he was not persuaded that the South Downs merited national park status. We decided to invite him and his chairman to visit the South Downs but without any real hope that they would change their opinion in the near future.

The Commission was not expected to make its formal position known on the best option for the future of the South Downs until the next AONB Forum meeting on 24 April 1991. In practise, the decision had already been taken at the Commission's Board meeting on 4 April which described the establishment of the Sussex Downs Conservation Board as *'an exciting possibility to achieve the objectives set out in the AONB policy statement published in July 1990.'*

On the day of the Forum, we released a press statement that drew attention to the differences between what was proposed for the governance of national parks in the Edwards Review and what was proposed for the South Downs. We described the latter as: *'a botch that won't work'* and urged *'the Commission to stop pressing ahead with indecent haste, thereby failing to take into account the recommendations of the Nation Parks Review Panel.'*

The Forum decided to support a Sussex Downs Conservation Board, but also agreed to keep under review the question of national park status. The Countryside Commission's Regional Officer tabled a statement saying that it would be some time before the Government responded to the Edwards Review report. There might be an exercise to identify new national parks depending on the outcome of that response. He explained that this process, if implemented, would take several years to come to fruition.

Afterwards we wrote to the West Sussex County Council secretary, re-affirming that we were convinced that a national park authority was the best solution. We said that we did not support the draft proposals for a conservation board because they were extremely weak and that we shared the aspirations of the Countryside Commission, which expressed hope that there would be a single unit for planning and management purposes. We would strive to ensure that a conservation board had its powers strengthened and its functions expanded. Our Campaign now represented a wide body of national and local organisations and if it would be helpful, we would offer a representative to be on the working party and to help strengthen the management machinery for the South Downs. No response was forthcoming and at our 2 July meeting we noted that a six-man working party set up by the Forum had already met. Richard Reed, the chair of the Society of Sussex Downsmen, had been appointed to represent 'conservation bodies'.

After all our strenuous attempts to have a South Downs National Park created had failed, and plans for the Conservation Board were clearly gaining momentum, we needed to reassess the situation. We agreed that our prime object remained the achievement of national park status for the South Downs. To stand aside from the Conservation Board would, however, risk isolation and ineffectiveness. Member organizations should therefore lend support to the setting up of a board, whilst constructively criticizing its inherent weaknesses. But we should not abandon our efforts to sway the Countryside Commission and Ministers and we should continue to work with others who favoured national park status.

It was at this point that we first began to move away from the concept of a 'chalk-only' national park. At this stage it was largely a pragmatic decision: if a conservation board was to cover the whole of the AONB, why not a national park authority to cover that same wider area? But it is fair to say that it took time for us to fully appreciate the threats to the Western Weald, particularly from weak planning policies, or the benefits that would accrue from including this area within a national park.

On 10 September we released another press release:

'Nearly three years ago, East and West Sussex County Councils pointed out the urgent need for co-ordinated action to protect the South Downs: to the need

for more money for essential work conservation work and visitor management; to the need for an overall approach to planning and development control; to the need to ensure the livelihood of downland communities and the viability of downland farming.

The Counties rejected pressing for national park status, notwithstanding 75% long term funding from the Countryside Commission, as they felt all the thirteen local councils might be more likely to work together in a conservation board with only the possibility of 50% short term funding for six years.

Even this temporary option would appear to be fading away as a result of disagreements between thirteen county, district and borough councils as to what to do, who pays what and who has the power.

The members of the Campaign still hold the view that the only right solution is national park status which would bring not just resources, but increased conservation powers to the South Downs. They are willing to support the Conservation Board as an interim step towards national park status but only if it offers a prospect of resolute action.'

This time we caught the imagination of the media. Both BBC TV (South) and TVS covered the story and there was also correspondence in The Times culminating in a letter signed by Fiona Reynolds on behalf of CPRE and Amanda Nobbs on behalf of CNP. We were also delighted by a BBC documentary called 'Division on the Downs'. Thirty minutes of exposure that helped to raise awareness of the issues with the wider public.

Amanda invited Sir John Johnson, chairman of the Countryside Commission, and a keen rambler, to visit the South Downs. In no time it was suggested from within the Commission that as many as thirty members and staff might be interested in visiting the Downs for half a day. We were very uneasy: might this lead to the rejection of the Downs again based on a half day visit? Eventually Sir John Johnson came accompanied only by the Regional Officer.

A further letter to the Parliamentary Secretary at the Department of the Environment at last elicited a response which merely referred us to the Countryside Commission! In doing so the Minister observed that the Commission, in designating the Sussex Downs an AONB, had already concluded that the area did not meet the criteria for national park status. Our second year thus ended in frustration, but the Campaign had established a firm base. We were determined to press on despite the fact that the Conservation Board was now poised to be established in the coming spring.

CHAPTER FIVE

MARKING TIME
1992-1997

With the Conservation Board about to be established, on 8 February 1992 I wrote once more to Sir John Johnson setting out the case for a South Downs National Park. I reproduce my letter here as it set out succinctly our case.

'The establishment of the Conservation Board is a useful start. But it is deficient in several respects. Its limited ability to influence planning and development control decisions does not measure up to the tremendous pressures on the area. We view the Board as a first step on the way to full national park status under the kind of independent authority already endorsed by the Government.

The South Downs were recommended by Hobhouse (1947) on the grounds of both scenic qualities and recreation value. However, the National Parks Commission concluded in 1956 that, while the area had great natural beauty, its recreational value had been greatly reduced by extensive cultivation.

In practice, the Downs have since proved themselves to be a major recreational resource catering for a variety of informal pursuits. Access is facilitated by an extensive network of paths which, in the intervening years have been improved by a standard second to none. The 'sweeping views' referred to by Hobhouse are still there to be enjoyed as are the areas of 'surprising solitude' on which he remarked.

There has been a substantial increase in the amount of open access land as a result of acquisition by public bodies, e.g. Wolstonbury Hill, Frog Firle and Seven Sisters Country Park.

Access to wooded areas has been improved because of practical implementation by the Forestry Commission of its policy of allowing visitors to wander at will.

The Environment Sensitive Areas scheme is encouraging the conversion of arable grassland with future potential for access. The recent major strengthening of the scheme should accelerate the process. Countryside Stewardship could be employed to similar effect.

Conservation has been enhanced by a continuing process of designation of nature reserves and SSSIs, the scheduling of archaeological sites as Ancient Monuments and the listing of Downland buildings and other structures as being of special architectural and historic interest.

All these factors point to the South Downs as satisfying the dual criteria for national park status recently reaffirmed by the Commission as being eminently suited to the kind of 'quiet enjoyment' consistent with the objective of conservation.

Recreational use is substantial and still growing because of greater mobility, improved access and population growth in the south-east. The South Downs Way has increased national awareness.

In practice the South Downs already function as a national park. They are currently denied the administration, powers and resources necessary to sustain that role.'

He replied that, now that the Government had made its response to the Edwards Review, the Commission would be addressing the question of new national parks, but in the meantime its immediate priority was to see the Sussex Downs Conservation Board established and successful.

The Sussex Downs Conservation Board came into existence on 3 April 1992. It had a budget of £1M per annum, half of which was provided by the Countryside Commission and the rest by the local authorities. Its membership comprised six councillors each from East Sussex County Council and West Sussex County Council, twelve from the 11 District Councils (Chichester had two because of its large area in the AONB) and twelve nominated by the Countryside Commission to represent local farming, conservation, recreation, economic and community interests. Not all those nominated for the Board by our member organisations were accepted, but we were fortunate that two members of our committee, Peter Brandon and Brian Short, joined it as nominees of the Countryside Commission.

Another nominee of the Commission was a certain Len Clark, unknown to most of us at that time. This wise and highly articulate gentleman had a deep knowledge and love of national parks which began when he took a girlfriend on their first date to the second reading of the National Parks and Access to the Countryside Bill in 1949. Despite this somewhat novel approach to courting, they married and remained a devoted couple until his wife's death. He was to become a prominent and much-loved member of our Campaign from 1996.

Thus, the governance and management arrangements in the Sussex Downs

AONB seemed to be resolved for at least the next few years.

Things were also becoming more settled in the East Hampshire AONB. As early as 1968 Hampshire County Council had drawn up a range of policies relating to countryside management. Unfortunately, the local government reforms in the 1970s meant that little action had been taken.

Nevertheless, what was happening in the Sussex Downs AONB in the late 1980s attracted the attention of Hampshire County Council, who attended the Sussex Downs AONB Forum as observers from 1988 onwards. Also aware of happenings over the border, East Hampshire District Council took the initiative and held discussions with the Countryside Commission, Hampshire County Council and Winchester City Council with the idea of setting up a Joint Advisory Board and appointing an AONB Officer to manage the East Hampshire AONB. Hampshire County Council expressed concern about the financial implications and were sceptical of the idea of appointing an AONB officer who 'might act independently from the local authorities and might challenge them'. However, by the early 1990s an AONB officer had been appointed who was managed on a day-to-day basis by the Countryside Manager at Hampshire County Council. Her work programme, funding requirements and so forth were overseen by a steering group comprising the four funding partners and was chaired by the County Council. A Joint Advisory Committee, essentially a forum of interested local groups, chaired by a Hampshire County Council cabinet member, was also set up. The AONB officer became a non-statutory consultee on planning.

Just six days after the Sussex Downs Conservation Board became operational, the Conservatives won a general election victory. In their manifesto they had declared their commitment to national parks and to independent national park authorities:

'*All national park authorities will become independent boards, which will make it easier for them to carry out their tasks effectively. The New Forest will be given statutory status which will give it as great a level of protection as any national park. Consideration will be given to other areas.*'

However, the promised 'Green Bill', to include accelerated and flexible procedures for national park designations, failed to find a place in the Queen's speech. Progress, we learnt, was unlikely before 1993/94. The undertaking to set up independent authorities for national parks seemed likely to slip away even further to 1995. There was little we could do.

The Countryside Commission's Position Statement (August 1992) affirmed that: '*It believes that a higher priority, for the next few years, is to consolidate the position of existing national parks, to secure the new status for the New Forest, and to strengthen the planning and management of Areas of Outstanding Natural Beauty.'*

In July 1993, following the latest county council elections, we wrote to the newly elected councillors across the three counties highlighting the benefits of having a national park as opposed to the existing Conservation Board or the arrangements in Hampshire. We received few responses: most thought that the question of national park status was premature, and that in Sussex the Conservation Board should first be given a trial. At least they realised we had not gone away, and the issue remained a live one.

The following year, 1994, brought another concern regarding the Conservation Board, but this time it was an existential threat. The Local Government Commission consulted on a re-structure of local authorities that could have had significant repercussions on the make-up of the Conservation Board and which, at the very least, could have led to a hiatus in collaborative arrangements to conserve the Sussex Downs. We were not alone in our concerns: they were shared by the Board itself and they, like us, sent a robust paper to the Commission in response. The eventual outcome was the setting up of a unitary authority for Brighton and Hove, but this had little impact on the arrangements that were in place.

By 1995 there were murmurings from the Countryside Commission about the performance of the Conservation Board. They were concerned, we were told, about the delay in the preparation of a long-term Management Plan and were 'unimpressed' by what they regarded as its lack of imagination and flair. These concerns led to a meeting between the Commission nominees on the Board and Michael Dower, the Director General of the Commission, in which he indicated he had an open mind on the future of the Board. He made clear that the Commission would be looking for evidence of added value over and above the previous arrangements that would satisfy them that the additional money provided by the Commission had been well spent. Our source was Pat Leonard, a new recruit to the Conservation Board, who had been a member of our executive for just over a year. Pat had recently become a trustee of the Sussex Wildlife Trust. He had been an assistant director of the Countryside Commission and then head of rural development in the Department of the Environment. He was also one of the leading experts on fungi in the UK. He brought great weight and knowledge to our deliberations for a number of years before emigrating to the southern hemisphere.

We were delighted when the promised 'green' legislation finally surfaced in 1995 and contained significant provisions regarding national parks. The Environment Act set up national park authorities with full planning powers and included revised purposes for national park designation. The revised purposes were:
1. *To conserve and enhance the natural beauty, wildlife, and cultural heritage of the national parks; and*
2. *To promote opportunities for the understanding and enjoyment of the special qualities (of the national parks) by the public.*

The addition of a third purpose, to promote the economic and social development of communities within the national parks, as recommended by the Edwards Report, was rejected. However, a duty was imposed on national park authorities 'to seek to foster the economic and social well-being of local communities within the national park...': a parallel duty was imposed on relevant authorities to have regard to national park purposes.

The Environment Act also enshrined 'The Sandford Principle' in legislation which, in effect, meant that when there was an irreconcilable clash between conservation and recreation the former purpose had priority. All of this was excellent news, though the promised streamlining of the designation process for new national parks was not apparent on the face of the Act.

Meanwhile, we had a serious local problem to worry about. We were alerted to the fact that Brighton & Hove City Council were proposing to sell all the land outside or predominantly outside its boundaries. Eight farms, totalling 11,000 acres, were involved, most of the council's land holdings. This was truly shocking given the history of the land purchases in the 1930s and the reasons for them. There was a vigorous campaign to prevent this by a group set up by local residents known as Keep Our Downs Public, in which a few member organisations of the Campaign were involved. The energy of this local campaign generated considerable public support and media attention, demonstrating the strong connection that many residents in the Brighton area had for the Downs. It also helped highlight how vulnerable the Downs were and that existing provisions for their protection were not adequate. It was a very successful campaign. The Council was persuaded to withdraw the sale of the vast majority of the land in question. Crucially, it managed to convince the then Labour Leader of the Council, Steve Bassam (later Lord Bassam), to support a national park in the South Downs.

Despite the success of Keep Our Downs Public, I have to admit that, for those of us involved in the Campaign at the time, it raised serious concerns. Its strong combative media presence was a method of campaigning that we were neither used to nor felt comfortable with. It also led to the Friends of the Earth calling for a *'Sub-regional campaign for a South Downs National Park style authority'*, that cut across our own actions and policies. Uncomfortable though this all was, there is little doubt it re-invigorated the debate about the future of the Downs in that part of Sussex, tapping into concerns about the threats from development and intensive farming.

Consequently, there was a serious debate about how we could work together with the Brighton campaigners to achieve our common aim. Fortunately, we came together in an amicable agreement that, I believe, benefitted all concerned. There is no doubt in my mind that a more vigorous media presence helped to enhance the profile of the Campaign and persuade many more people to support its aims. This in turn was important in the times ahead as it raised our profile in the minds of those influential in the Labour Party, which was a cornerstone of the ensuing campaign.

Later, in 1996, the Sussex Downs Conservation Board launched a public consultation regarding its future for submission to the Countryside Commission. It was necessary for them to do this since the then current experimental arrangements only extended to March 1998 and Countryside Commission funding, which had made a material difference to the management of the Downs, would cease at that point. The consultation set out the various options including a permanent conservation board which, we believed, would require new legislation; a special purpose authority on the lines of the Broads Authority or a national park under the 1949 Act. The latter option was split into two alternatives: a national park encompassing the area of the AONB or one constrained to the chalk. Whatever path was taken we wanted the Hampshire section of the Downs to be included. We decided that it would be premature to take a view on any preferred form of future organisation. Our submission concluded:

'It is essential for the future of the Downs that a clear decision is taken now on their future administration. There is a strong case for a Statutory Body to protect the South Downs with the necessary powers and finance necessary to deliver its objectives.

The options best available to deliver this appear to be a special authority of national park status modelled on the Broads Authority or a national park. It seems to us to be premature to select one and rule out the other. Both merit the consideration of the Board.'

The result of the consultation, published in the July of that year, showed overwhelming support from the local authorities and landowners for the continuance of the Conservation Board with its present powers. East Sussex County Council supported a special statutory authority, as did the National Trust, CPRE Sussex and the Society of Sussex Downsmen. Support for the national park option was almost entirely confined to our supporting organisations.

This consultation brought to us another stalwart of the Campaign. John Templeton became involved thanks to Len Clark, who he knew through the Youth Hostels Association. Born in Chichester, John had qualified as a planner, finishing his career in Ealing. He had been the YHA's representative on CNP and had been a member of the consultative panel campaigning for a New Forest National Park. John's passion for protected landscapes, together with his continued interest in his hometown of Chichester, inevitably led him in our direction and he became the YHA's representative on our committee. A quiet unassuming man, he was a great asset to us and his planning experience invaluable. A great enthusiast for the cause, he remained actively involved until the end of the Campaign, which he maintained to be the most enjoyable and stimulating time of his life.

In September 1996 we received a depressing letter from Richard Symonds, now chairman of the Countryside Commission, that said the Commission had concluded that the Sussex Downs Conservation Board had been a success and they would like to see a new conservation board established which would also embrace the East Hampshire AONB. They rejected the national park option, noting there was widespread opposition, especially from the local authorities, and approval in the current climate would be unlikely.

In contrast, we had a very warm letter from Lord Nathan, chairman of the Sussex Downs Conservation Board, thanking us for such powerful support and great encouragement for the South Downs.

Throughout this period, we were relentlessly having to counter numerous statements and letters in the press, particularly from West Sussex County Council, which we believed to be at best misleading or simply not true. We were well aware that the term 'national park' denoted for some centralised decision making or even nationalisation of the land. It was a problem that remained for many years to come despite the Environment Act setting out that three-quarters of the membership of national park authorities should be county, district or parish councillors from the local area. This should have reassured opponents that the more centralised approach favoured by Dower was outdated but this was not to be. For example, we received a letter from Ben Perkins on behalf of the Society of Sussex Downsmen, in which he stated that their Council

'were already getting flak from their members who felt they had been sold out to the national park lobby, based on the Times letter. Their official position remained one of opposition to national park designation, though not to the idea of equivalent status'.

Ben Perkins was a charming man. A GP in Hove, he eventually had 14 books on walking published, including one on the Sussex Border Path, a 150-mile footpath route around the borders of East and West Sussex, linking Emsworth with Rye, which he designed and co-wrote. He contributed a detailed walking column to The Argus every fortnight for 28 years until 2012, and he aimed to draw ramblers' attention to lesser-known paths and walks with a good variety of landscapes. He joined our Executive as an 'observer' and his knowledge and wisdom was much appreciated as he trod a somewhat delicate path, given the divisions within the Sussex Downsmen.

Despite all our efforts over six years, it was only too clear that the Campaign had a long and formidable fight on its hands if it was going to achieve its objective of a South Downs National Park.

CHAPTER SIX

CHANGE AFOOT

1997 was going to be a General Election year. This stirred us into action, and we produced a comprehensive paper, based on an original draft by Phil Belden, entitled 'The South Downs: Securing their future' calling for a national park or equivalent status. In it we highlighted the special qualities of the South Downs, its threats and opportunities and our vision. We countered the myths and misconceptions such as: 'that the Downs are protected already', 'The Downs are not wild, they don't qualify for national park status,' 'A national park would attract too many visitors,' 'National park status would take away local control' and 'A national park would create an additional bureaucracy.'

This we distributed to national and local politicians, parliamentary election candidates, farmers and landowners, the Countryside Commission, Department of the Environment, English Nature, and voluntary organisations. It elicited many responses as well as acknowledgements from the offices of the Prime Minister and the Liberal leader Paddy Ashdown. Chris Todd, who was to become a key person in our campaign, handed a copy to the Labour Leader Tony Blair during the latter's visit to the University of Sussex.

The most extraordinary response was in a letter in the Chichester, Midhurst and Petworth Observer papers which read '...*the motley crew of professional special interest managers, who have formed the South Downs Campaign Group, have their own interests, and not the well-being of the Downland residents in mind. National parks have tourist and travel objectives in mind'... 'Our priceless heritage will be afflicted by a kind of blight which invariably follows on from control by unaccountable managers in far-away office blocks. I urge readers not to allow this self-interested and self-appointed group of busy-bodying lobbyists to get their foot in the door.*'

In a report to the executive committee of the Sussex Downs Conservation Board, the clerk wrote a long and detailed response to our paper, which he acknowledged as being a cogent and well-written case for national park status and a helpful contribution to the debate. He set out the now familiar objections

© *Tim Rice FOE*
'Unploughing' Offham Down Site of Special Scientific Interest

to a South Downs National Park and concluded that the 1995 Act did not provide the best way forward for the Downs. As an alternative he drafted a clause for possible inclusion in a future bill on AONBs which would meet the Conservation Board's requirements. He kindly sent us his draft clause which we strongly challenged.

Meanwhile happenings on the ground near Brighton starkly exposed the weaknesses in the environmental schemes then in place. At New Erringham Farm, the farmer exercised the break agreement in the Environmentally Sensitive Area scheme and ploughed up five years of subsidies at a stroke. Initially there was little media interest in the story but a short piece in the Daily Telegraph changed all that and suddenly there was a media rush for photographs of the damaged field.

But this was as nothing to what happened barely a week later. A farmer began to plough up irreplaceable ancient chalk grassland on Offham Down that was nationally protected as a Site of Special Scientific Interest. This came about because of the absurdities of the subsidy regime that meant that an individual could get more than £300 per acre for ploughing the land and only £14 or £15 per acre for conserving it. English Nature were also reluctant to enter into an agreement which would have prevented the ploughing. Alerted to what was happening, activists from Brighton and Lewes vowed to stop the damage. Dave Bangs bravely stood in front of the tractor in a dangerous game of cat and mouse to prevent further damage. Eventually, the driver gave up and left the field. A camp was promptly set up on the site to protect it. Chris Todd later distributed flyers giving instructions on how to 'unplough' the damaged land and over 250 volunteers responded, turning the turfs back green side up. The story hit the national press and alerted politicians to what was happening, and soon two Labour prospective Parliamentary candidates for Brighton visited the site with Michael Meacher followed soon after by Norman Baker, the LibDem candidate for Lewes.

This shocking event happened just three weeks before the General Election. Cross-party working ensured that John Gummer, the Secretary of State for the Environment, issued only his fourth ever stop notice to prevent further harm. The General Election was held on Thursday 1 May 1997 with Labour ending its eighteen-year spell in opposition with a landslide victory. David Lepper, the newly elected Labour MP for Brighton Pavilion, remembered the event very well:

'Back in Brighton in 1997 to back my parliamentary election campaign, Michael Meacher asked me to go with him when he accepted an invitation from South Downs campaigners to visit Offham and see what was happening to the SSSI there.

It was the sight of that scarred slope, the attempts by campaigners to reinstate it and the arguments they put that day which prompted my questioning of the adequacy of that existing protection. As I told The Argus we needed "A radical rethink about how we protect our countryside."

..After the General Election Meacher, now Environment Minister in the new government, influenced English Nature to negotiate a new management agreement with the farmer for another SSSI he owned – Offham Marshes. And he told his Departmental officials to begin discussions with "interested parties" on action needed to give greater protection to SSSIs, some 200-300 of which were lost or damaged every year.

For me Offham Down began more than my own involvement in the process which eventually led to the designation of the South Downs as a national park. While that remained the local focus it was also the start of my interest in strengthening the protection needed by all threatened wildlife sites.'

With a new Government in place, we saw an opportunity to re-ignite the debate on the future arrangements for protecting the South Downs. This was becoming increasingly urgent as the deadline of end of March 1998 was approaching fast and there was, as yet, no agreed way forward on what was to replace the experimental Conservation Board. Therefore, shortly after the General Election we wrote to Michael Meacher, urging him to move quickly on the future arrangements for the South Downs and to consider national park status rather than allow its future to be caught up in a protracted debate on the future of AONBs in general.

David Lepper well understood the urgency:

'The South Downs issue was urgent because the Sussex Downs Conservation Board's (SDCB) remit was to end in April 1998 and, even if nothing else was settled, local councils needed to know what the contribution from their budgets to its successor would be. And so in June 1997 Brighton and Hove MPs Des Turner, Ivor Caplin and I began what was to be a 10 year conversation with a succession of secretaries of state and ministers starting with Michael Meacher and ending with Hilary Benn and Huw Irranca-Davies, often involving representatives of the South Downs Campaign which had been formed in 1990, Friends of the Earth (FoE) and others. That time span shows the complexity of the arrangements which would be needed.

For us MPs the first success in July 1997 was Meacher's 3- year extension of the SDCB's life and a statement that he would "like to see a long-term solution in place before the end... of that interim period." The Countryside Commission (CC) was to consider long term options.'

The Countryside Commission proposals for the short term were made known early that month: beyond March 1998 it would continue to fund the Conservation Board for another three years, to the tune of just over a million pounds in total. This was considerably less money than had previously been provided and we heard that it had left the morale of the Board's staff at rock bottom. The terms attached to the money were described as unworkable. Though the Board had been established to oversee the management of the Downs, the new money was not to be spent on funding the strategic core but to be used for 'countryside projects'.

We were not alone in our disappointment at this news. On 4 July 1997 Howard Flight, then the Conservative MP for Arundel and South Downs, raised the subject of the Sussex Downs Conservation Board and its' funding in a brief Adjournment Debate in the House of Commons. In doing so he somewhat breathtakingly claimed that it was the Conservation Board who had stopped the destruction on Offham Down and had succeeded in putting back the sods of earth. In response, Angela Eagle, then a Parliamentary Under Secretary of State at the Department of the Environment, Transport and the Regions stated that, as promised before the election, the Government intended to institute a consultation process on the best way to manage the Downs in the medium and longer term. The Countryside Commission would issue a separate consultation in September about a possible system for managing and funding AONBs.

This debate in the Commons was followed by a rather longer one in the House of Lords on 22 July. It was instigated by Viscount Mersey who, despite his title, had been born and bred at Bignor, site of the renowned Roman Villa. He praised the work of the Conservation Board and made the case for its continuation. In this he was supported by a number of other peers from the area, most notably the new chair of the Conservation Board, Lord Renton, who stated:

'I see no sign in Sussex of great support for national park status… Farmers and landowners are against it. Local Authorities are against it.'

Peers with connections to CPRE, CNP and the Ramblers Association spoke in favour of a national park solution. Less supportive was the president of the Society of Sussex Downsmen, Lord Addison, who suggested that the Conservation Board, *'or something rather similar with marginally but not greatly increased powers'* was needed. The response for the Government reiterated that it was seeking a long-term solution and expressed the hope that this would be in place by the third year of the extended funding period.

In the meantime, a group of us joined a meeting in Cheltenham with the Countryside Commission to discuss the Government's commitment to designate more national parks, the specific cases of the South Downs and the New Forest, and the Countryside Commission's forthcoming consultation on the future management of AONBs. It was clear that the Countryside Commission's limited resources were much stretched, and they were horrified at the growing gap in funding between national parks and AONBs. They implied that they did not want the South Downs to be given priority at the expense of other AONBs. We were therefore concerned that the Countryside Commission's current policy was firmly against granting national park status to the South Downs.

Consequently, we wrote to Angela Eagle *'to seek assurance that the consultation process you propose to arrange regarding the future of the South Downs will, to quote Lord Addison in the debate in the House of Lords last week, "genuinely explore all options for the South Downs, including that of awarding them national park status."* In response we were assured by Angela Eagle that *'The Commission propose to run the South Downs consultation from its London office, separately to the wider AONB consultation exercise which will be run from the Cheltenham office.'*

We were determined that the national park option should get a fair hearing this time.

CHAPTER SEVEN

PROGRESS

The Campaign now entered a crucial phase. We had a new government that seemed open-minded about the future arrangements for the South Downs and was prepared to keep open the option of a national park. However, the process was in the hands of the Countryside Commission: they alone could recommend whether to designate a new national park or not. At the end of the process, the final decision rested with the Secretary of State, but he had to act in a quasi-judicial capacity. Politicians therefore had to walk a delicate path and not seek to impose their will on officers in the Commission. All the same, they could influence the tenor of the debate and their support was therefore potentially of enormous value.

It had been clear to us for some time that the Countryside Commission were far from persuaded of the merits of our case. They seemed to be more interested in improving the lot of the AONBs more generally than in designating new national parks. And they seemed nervous of the strength of opposition to a South Downs National Park from local councils and landowners. I think it is no exaggeration to say that had the Campaign not existed to keep up the pressure there would be no South Downs National Park today.

At the beginning of September 1997, the Countryside Commission launched the promised consultation on the future of the 37 nationally important English landscapes designated as AONBs. So far as the South Downs were concerned, the press statement followed the line taken by Angela Eagle in the House of Commons debate:

'Environment Minister Michael Meacher has said that he hopes that long term management arrangements for the whole of the South Downs will be in place before the end of the current funding of the Sussex Downs Conservation Board. To that end he asked the Commission to look at the needs of the South Downs, so that he is best able to decide the most appropriate type of organisation we need.

All options are to be considered, including the possibility of designating a new national park. The examinations of the options already carried out by the

Sussex Downs Conservation Board provides a good start. The Commission will shortly hold a conference and consultation process to identify those needs.'

So far so good.

However, one of our executive committee members found buried in the detail of the consultation document a significant statement:

'Whilst the Countryside Commission's consultation document was addressed to the generality of AONBs, it contains important qualifications... that Statutory Conservation Boards would apply only to AONBs which have a strong case for special consideration. This approach would provide continuity and long-term confidence for the Sussex Downs Conservation Board in particular.'

This recommendation distinctly resembled the approach which the clerk to the Conservation Board had urged upon his Board. We were most concerned that, once again, we were facing a fait-accompli.

It was against this background that we prepared for the promised South Downs Conference. Ben Perkins had kindly offered the Downsmen's office in Hove as our regular meeting place. This was a convenient venue which we used for several years before moving to the Friends Meeting Hall in Chichester in 2000. Our gathering in Hove on 29 August was the first time Chris Todd attended, representing Friends of the Earth. We resolved to ensure that that this consultation and the planned conference on the future of the South Downs should be as thorough and inclusive as possible, unlike the previous one on the Conservation Board, when just forty individual members of the public had responded.

The conference took place in November 1997. It was chaired by Professor Chris Baines, a landscape architect, and a national leader in the environmental voluntary sector. It was an important event which encapsulated the diverse views on the future management of the Downs current at the time and so the main thrust of the interventions is worthy of inclusion here.

The spokesman for the Countryside Commission set the scene. He began by asking a series of questions about the future management of AONBs generally in terms of their governance, financing, powers and responsibilities. He then moved on to outline four possible options for the South Downs: the first two related to the continuation of the AONBs, either under broadly the existing arrangements or through improved legislation. The other options for consideration were either a national park under the 1949 Act or a tailor-made national park solution, which would require a Private Bill promoted by the local authorities and backed by Government.

Lord Renton, speaking as Chairman of the Sussex Downs Conservation Board, and in favour of its retention, said that the Board was recognised by the Countryside Commission, local authorities and voluntary bodies as an effective voice, however two long term issues needed to be resolved. The first was the Board's role in planning. Although the existing arrangements had not yet been fully tested, its planning role needed to be strengthened, but this was not best achieved by the Board becoming the sole planning authority for the area. The second was the basis for long-term arrangements. Local authorities bore the burden of financing the Board, but they could not be expected to do so permanently given the South Downs were a national attraction.

Merrick Denton-Thompson, a senior officer from Hampshire County Council, was the next to speak. He concentrated more on the strategic challenges facing the area rather than administrative solutions though he expressed himself not in favour of a body that was a statutory consultee on planning. The main concern he expressed was the impact of agriculture on the area, the neglect of woodlands and the loss of wildlife habitats. In summary he said that any new administration should make it a condition of any public investment that sound environmental objectives were strictly adhered to that would conserve and enhance the quality of the environment. Somewhat ambitiously, he suggested the South Downs should become a pilot which, in the long term, could benefit the wider reforms of the Common Agricultural Policy. He concluded by saying *'Do not be constrained by existing models such as the Sussex Downs Conservation Board or for that matter the East Hants Joint Advisory Committee in Hampshire'*.

Ian Elliott from West Sussex County Council spoke for all the Sussex local authorities save for Brighton & Hove. He favoured the retention of the Sussex Downs Conservation Board and said the local authorities were committed to providing two thirds of the funding for it. Although they would like to give the Board a right to be heard on planning issues, they did not want to have their planning control taken away. They did not support the idea that the Environment Act was flexible enough to safeguard the local authorities' position. Secure funding was needed from both central and local government.

Alastair Nugent, a farmer and landowner, said that he wanted to see a new era of co-operation between those who lived and worked on the Downs and visitors who were their guests, with respect for county traditions and creative dialogue.

Trevor Cherrett, for Sussex Rural Community Council, said they had conducted a study of rural community needs. It found a low level of village services; a shortage of affordable housing; declining agricultural employment and elderly people increasing at a greater rate than elsewhere in Sussex. Future arrangements were required which represented and were accountable to different interests: arrangements that could create a vision of a sustainable cultural landscape and develop a coherent strategy which attracted widespread support. A rounded view of a cultural landscape was needed, which could bring together appropriate bodies; a partnership and inclusive approach which gave local committees a full say and full share in a future vision for the South Downs.

Chris Todd spoke on behalf of the Friends of the Earth. He emphasised that there was the need for a more pro-active public consultation. Friends of the Earth had created much publicity about this conference, which would otherwise have taken place with little public awareness. It was a rather exclusive event with most of the attendees not representing the majority of South Downs users who were urban based. Better ways needed to be found to engage more people in a thoroughly inclusive consultation.

I spoke on behalf of the South Downs Campaign: it was an important moment since it was the first opportunity to present our arguments comprehensively in public. Conscious of this, my presentation was the product of many hours of work and much agonising by several of our committee members. I began with the history of the conservation efforts on the South Downs and the area's clear value to the public. I set out the weaknesses we saw in the conservation board model: in particular, those regarding the fragmentation of planning where there were in play three county strategic plans and twelve other planning authorities. I dismissed some of the myths about a national park authority: in statute it would be a special purpose local authority not a quango. I explained the advantages of having a body that could play a strategic role in environmental matters and in managing agro-environmental schemes. All in all, I think I made a strong and persuasive case for a national park.

It seemed to me that we had rattled those favouring a conservation board. On my arrival, I was just about to be interviewed by a TV reporter when West Sussex County Council's Planning Officer, said somewhat patronizingly I thought: 'Hello, are you having your moment of glory?' And, during the luncheon interval, Lord Renton accused me of lying. He had in fact misheard what I had said: I had said that the Sussex Downs Conservation Board's officers were also answerable to West Sussex County Council: he had missed the important 'also'.

In the debate after the presentations, we were told that my quote that national planning guidance affirmed that national parks are afforded the highest degree of protection was incorrect even though the Countryside Commission's own consultation paper stated that '*AONBs cover areas of high scenic quality and, in landscape terms, they are of equal status as national parks. As such they are given special protection in the planning system, but not as strong as in national parks.*' (emphasis added)

I had a wry smile when I read in the Countryside Commission's written summary of the conference that one of our most outspoken opponents had said that he was '*impressed by the sincerity of Robin Crane's address*'.

Following the conference, as the consultation proceeded, it became apparent that two issues dominated the thinking of many of the affected local authorities in Sussex. These were planning and visitor numbers.

At that time county councils were responsible for strategic planning. They drew up the minerals and waste plans and determined the framework for development, including the number of houses to be built in each district or borough area. It was then for those district or borough councils to draw up the policies that would set out how their own area would develop and the location for housing and economic activity. Planning applications were then determined against this policy framework. This was an important part of the work of these councils. Understandably from their perspective, the local authorities did not want to cede planning to any new body.

This was the conundrum. It had been amply demonstrated in national parks elsewhere in the country that, to fulfil the purposes of designation, it was necessary to have a single planning authority for the whole national park area, hence the findings of the Edwards Report. This was certainly the case in the South Downs given the number of local authorities concerned and the lack of consistency in their approach to the protected areas. But if such an authority came into being, then there risked being fragmentation of planning within the affected districts and boroughs. Some districts, like Mid Sussex, had only a small fraction of their district in the AONB and this lay at the southernmost extremity of their area. But, in the case of Chichester District Council, where something like 70% of their area was in the Sussex Downs AONB, a national park with planning powers risked leaving them with only the area to the south of the Downs and an 'outlier' in the north-east potentially under their planning control, despite their retaining other responsibilities for the whole of the district area.

The councils claimed that a national park authority with planning powers would cost £1M more than the existing system. We countered by emphasising the need for a single voice on planning to replace the fragmented system in place.

We also queried the price tag, arguing that as the number of applications would be the same, any additional costs falling to a national park authority would be matched by a reduction in the cost of planning to the local authorities. I think this is one of the issues on which, in retrospect, we were in error. It is certainly the case that a national park authority would be far better placed to consider the area as a whole and produce a sub-regional planning strategy worthy of one of the most important landscapes in the country. But, at the time, we did not appreciate the costs involved in putting that in place.

The local authorities also expressed concern about visitor pressure. There was no doubt about the popularity of the South Downs as a visitor destination, and the anxieties this raised for farmers and landowners, particularly those located near 'honey pots' or urban areas. The local authorities argued that national park authorities had, as part of their remit, tourism, and that national park status would increase visitor numbers. They also argued that as AONBs were designated solely for their natural beauty they were better able to concentrate on conservation. To give some flavour of the views then being expressed I reproduce just two quotes:

Chairman WSCC Countryside Committee:

'…. *Without local involvement, there is a great danger of the South Downs becoming little more than the national playground of the "Sunny South". The increased pressure will be intolerable; farmers will be forced on the defensive and be concerned only for their own interests. The creation of a national park would be a disaster – decisions would no longer be taken by local people properly elected by the local community, but by central government promoting "interests in recreation" in a way not required in present circumstances.*'

Chichester District Councillor:

'*National parks are managed with tourist and travel industry objects in mind.*'

'*What is needed is a clear statement that, whilst a national park authority would promote public access and enjoyment, it would have a statutory duty to ensure that this would not be done in any way that would be detrimental to the natural beauty, wildlife and cultural heritage.*'

These views were inaccurate on several counts: national park authorities are not tourism authorities, that role remains with the local authorities. It is also the case that the second purpose for national park designation talks of encouraging the understanding and quiet enjoyment of the special qualities of the area, thus placing an obligation on national park authorities to manage recreational pressure accordingly.

Finally, there is the Sandford Principle, which gives precedence to conservation over recreation should there be an irreconcilable clash between the two.

The one Sussex local authority to consult its people at this time was Brighton & Hove, an event well remembered by David Lepper:

'In Brighton and Hove the new City Council consulted residents – the only council to do so – and subsequently gave its backing to the national park option. The support was cross party including both the Labour administration and Conservative opposition leader Geoffrey Theobald and it was symbolically important. Not only because the cross-party support was at odds with that of other local councils. The Council still owned stretches of downland and its support reversed historic opposition which had been a factor in the exclusion of the South Downs from the national parks family following the Hobhouse Report.'

A number of us spoke at meetings we set up across Sussex and into Hampshire. An open meeting at Eastbourne chaired by David Dimbleby voted overwhelmingly for a national park. Elsewhere too, individuals and voluntary groups were in favour of a body with stronger powers. Our analysis suggested that support for a national park was highest in the east and diminished towards the west. We had to counter misinformation like that from a councillor at West Sussex County Council: *'A national park seemed to be almost like nationalisation of local resources. There are always people who want to take things away from local control and give it to some nebulous body which operated the levers at commanding heights. It would mean more bureaucracy, more rules and less freedom for the quiet appreciation of the Downs.'*

The publication of the Countryside Commission's draft 'Advice to Government' following the consultation led us finally to lose all confidence in the impartiality of the Commission. We were incensed that the Commission seemed to have ignored their undertaking at the opening of the consultation to base their decision on the quality of the arguments submitted. They simply produced a statistical analysis of the responses and we didn't even agree with that. Our own analysis suggested that most respondents wanted to see a new authority with a statutory role in the planning system. They wanted it to have all the powers and policies of a national park authority, though there was a push for development control decisions to rest with existing planning authorities. None of this was evident in the report. However, the report did acknowledge a growing enthusiasm for the inclusion of the whole of both the Sussex Downs AONB and East Hampshire AONB within the management of a single authority.

We were appalled to see in the paper to be presented to the Countryside Commissioners' Board meeting the recommendation that the most appropriate arrangement for the South Downs was as an 'enhanced AONB'. Officers advised the Board that: *'designation as a national park would not be right under the current law and policy and full national park authority powers are not necessary to meet the management needs identified through consultation.'* Embedded in the document was a blatantly untrue statement that *'the Edwards Report concluded that the New Forest was the only remaining area appropriate for national park designation'*. The Edwards Report had actually said: *'Given the time available to us to undertake our task, it was not possible to visit those areas suggested as new national parks...The possible areas for designation given to us include areas of nationally important landscapes in the lowlands or the coast of England which are under great pressure from recreation and development (for example the South Downs).'*

In anticipation of that Countryside Commission meeting, when the Board would be considering the future of AONBs and the South Downs, we had a letter published in The Times calling for a South Downs National Park that was signed by Chris Bonington (Council for National Parks), David Bellamy (Wildlife Trusts), Derek Hanson (YHA), Norma Johnston (Ramblers Association), Rodney Clegg (Open Spaces Society) and myself (South Downs Campaign).

Phil Belden, Alison Marshall, a recently appointed policy officer of CNP, and Len Clark attended the Countryside Commissioners' Board meeting in Leeds on 23 April 1998 where the Board approved their officers' recommendations on the South Downs. All three of our members were appalled by the manner in which the subject was debated. Len Clark sent me perhaps the most moderate account of the meeting:

'The officers reported that the area did not meet the criteria for national park status (extensive areas of wild countryside). There was also mention of local authority opposition. Commissioners laboured on the issue and repeatedly said it was difficult to resolve. It did not appear that any of the speakers had much direct knowledge of the Downs and rejection was led by a land agent from Northumberland! There was discussion on the inherent quality of the area but no discussion on planning issues other than officers' advice that this would cost £1M a year – though the consensus is for delegation of development control, which is the bulk of the work, back to local authorities. Nor was there much consideration given to the results of the recent consultation, which showed a stronger wish for a national park (or similar) than for other solutions. Altogether an unsatisfactory

discussion, in which there was a suspicion that the inbuilt interests of landowners against a national park was a motivator, though never overt.'

Phil Belden and Alison Marshall were less constrained in their report to our next Executive meeting when they expressed the view that the decision was *'based on false assumptions, misinformation, lack of information and lack of discussion.'*

In the Sussex Downs Conservation Board's press release Ian Elliott, now Vice Chairman of the Board, said: *'There is no good reason to lose all the goodwill in exchange for an imposed national park that would interfere in everyone's lives and for which neither greater protection nor certain funding is guaranteed.'*

A week later we had a meeting with an official at the Department for the Environment Transport and the Regions, to express our concerns about the Countryside Commission's decision. I was accompanied by Pat Leonard, Amanda Nobbs and Alison Marshall. We explained that our objective was to get the best solution for the Downs. We had come to the view through painstaking consideration of all the factors at our disposal that a national park was the option that would deliver what was needed most pragmatically.

We went on to say that the strength of support for a national park was impressive. It was our experience that, once all sides had been debated at public meetings, the majority were pro national park. It seemed to us that neither the staff nor the Commissioners were in touch with the feeling on the ground. We were particularly concerned that, apart from their chairman, none of the Commissioners who were making these crucial decisions had visited the South Downs. No evidence was produced to explain why the South Downs did not fit the criteria for a national park.

We expressed concern that, at their meeting, the Commissioners had not been told that 62% of respondents to the consultation favoured national park designation. They had not discussed national park criteria, nor why their paper stated that *'a national park authority will confer more than the required powers.'* We were also concerned that the Countryside Commission's proposals for an 'enhanced AONB' route would require new legislation that would depend on getting parliamentary time and thus delay the establishment of a permanent body.

Shortly afterwards we received a letter from the Countryside Commission's South-East Regional Officer brushing aside our objections to the published results of the consultation and reassuring us that the proposals that the Commission would be putting to Government would deliver the very best protection and management of the Downs. The arrangements would, through new legislation, allow for a permanent independent statutory conservation board, funded by Government and have considerably greater powers and responsibilities than the existing Board. It would strengthen the position of AONBs in public policy, give strong influence on planning and agriculture, bring a new duty of care to all public bodies and give local authorities a duty to pursue the purposes of the designation.

At much the same time I had a memo from the Director General of The Wildlife Trusts. He was in close touch with Michael Meacher. He had gathered that the minister was well aware of our campaign, '*but the gist of the advice from the Commission seemed to be that "the South Downs lot" want the protection that national park status offers, but that would bring thousands more visitors, that AONB status will give exactly the same protection as national parks but would make it easier to stop thousands more visitors and therefore be better for conservation of the Downs.*'

On 26 June 1998 the Countryside Commission formally released its glossy publication 'Protecting our Finest Countryside: Advice to Government'. In it they wrote that it '*considers that the South Downs do not meet the criteria for designation as a national park as presently defined and applied. Designation was rejected in the 1950s because the area was not appropriate for national park status. Although the area qualified as being of great natural beauty and was accessible to centres of population, its recreational value as a potential national park had been considerably reduced by the extensive cultivation of the Downland. It no longer had sufficient extensive tracts of open country, which provided opportunities for open air recreation.*'

'*If the South Downs were designated, the criteria for designation, confirmed in the Edwards Report (1991) would need to be changed. This would lead to demands for designation of other areas with similar qualities and recreational facilities e.g. the Chilterns and the Cotswolds. Consistent treatment would make it very hard for the Commission to resist such demands.*'

There were other reasons for rejecting national park status such as that '*the boundaries of a national park would have to be decided. Some areas of the existing South Downs AONBs would not now meet the present criteria, and there would be pressure to exclude part of the currently designated AONB (both in the West Weald and land on the fringes of Brighton conurbation might have to be excluded).*'

A month later we wrote to Michael Meacher enclosing our detailed response to the Countryside Commission's document. In it we emphasised that, in our view, the Countryside Commission had provided no evidence or recent research to back up their many assertions, in particular the conclusions that the area lacked the necessary quality or recreational experience. We evidenced the only recent survey carried out on recreational value which had been undertaken by Dave Bangs for a Freedom to Roam Group in response to a consultation on open access. We also pointed out the many inconsistencies in the Commission's document culminating in a statement that a national park designation process for the South Downs would be expensive, lengthy and difficult despite recommending precisely that same process should be applied to the New Forest.

I believe our challenge was well made. The Countryside Commission's assertions regarding the lack of recreational value did not bear close examination. I have already alluded to the Environmental Sensitive Area (ESA) scheme launched in the South Downs in the 1980s. Despite the weaknesses I described earlier, it had led to landscape-scale reversion of arable land to grassland on the Downs. In 1991 the Countryside Commission had launched another, complementary, agro-environmental scheme, the Countryside Stewardship Scheme (CSS), which was directed towards smaller-scale habitat restoration but, unlike the ESA scheme, it was not confined to the chalk Downs and could be used for such activities as lowland heath restoration and hedge planting in the Western Weald. The consequence was that by this time, despite all the inadequacies of these programmes, progress was being made in fixing the worst excesses of post-war agricultural practices in the area. Although we were not aware of it at the time, this had been reported in a research paper published in June 1997 and updated in April 1998. Moreover, though we were critical of the Conservation Board's lack of powers, particularly in relation to planning, we recognised that they had done a magnificent job in maintaining the rights of way to a high standard and their volunteers had also been engaged in activities such scrub clearance that helped maintain valued habitats. None of this was acknowledged in the Countryside Commission's publication.

Incensed as we were, we briefly flirted with the idea of applying for a judicial review of the Countryside Commission's decision. However, we were advised against this route, wisely, because if we had lost that would have almost certainly put an end to our Campaign.

Given the situation we were in, we decided that we must increase the pressure on Ministers. Vicki Elcoate was now the Director of CNP. Vicki had joined CNP in 1992 as Amanda Nobbs' deputy. Her previous career had been as a journalist with ITN lunchtime news when a video she produced for them called 'Battle for the Earth' persuaded her that she wanted to take a post graduate degree in environmental studies and move to the environmental sector. After Amanda left CNP, Vicki took her place and became a strong advocate for the South Downs.

She reflected on the challenges we faced at that time from CNP's perspective:

'The main problem was the attitude of the Countryside Commission, which opposed the cause seeking instead to gain equal status for AONBs and national parks. We were even threatened with having our funding withdrawn if we didn't take up the cause of AONBs and drop the South Downs campaign. This didn't cut any ice with our members who were firm in their resolve that the South Downs National Park was one of CNP's primary goals.

There was also a lot of opposition from some leading national park influencers who saw the South Downs as a potential watering down of what national parks were all about (not fitting the Dartmoor or Peak District mould). There was one memorable dinner when the Environment Minister Michael Meacher came as the special guest to discuss the issue. There was a lot of lively debate around the table from all points of view. Afterwards, perhaps because the table was large and people were loud and it was difficult to hear, he welcomed the enthusiasm of the gathering for the proposal. This was quite a relief to those of us campaigning for the South Downs!

Our role was to keep the spotlight on the issue and to engage public support across the country. It was an ambitious concept – to have a national park in the over-populated and developed south-east – and a wonderful thing if it could be achieved as it would deliver national park purposes across such a wide area that desperately needed better protection and investment in the creative solutions needed to balance development, environmental protection and public enjoyment and understanding. It was the long-term nature of the potential outcome that was so attractive and made it so worth fighting for.'

On 22 September 1998, Alan Meale, a Parliamentary Under-Secretary of State at the Department of the Environment, Transport and the Regions, working with Michael Meacher, visited the Downs to meet the Conservation Board. We arranged for Amanda Nobbs, Chris Todd and Dave Bangs to meet him on our behalf. Friends of the Earth organised a 'reception' for him at Brighton station to emphasise the huge public support for a South Downs National Park that, we suspected, was being underplayed by the Countryside Commission. Our subsequent meeting with him was succinct but useful. The Minister asked to have further information on Dave Bangs' research on 'open country' and agreed in principle to a further meeting with us.

We therefore wrote again to Michael Meacher enclosing an even better version of Dave Bangs' most revealing study of the South Downs. It had been strengthened by data provided by CNP which compared the pastoral land and woodland areas of the South Downs with all Britain's other national parks. We asserted that:

'Had the Countryside Commission performed a proper assessment, it would have been obliged to conclude that the South Downs did meet both the statutory criteria laid down in the National Parks and Access to the Countryside Act 1949 and the additional criteria of "openness and relative wildness" which the Countryside Commission had imposed.

A proper assessment by the Countryside Commission would have revealed as entirely false the statement that "even more land is under the plough (than in the 1950s)" indeed more open downland pastures, and thus more recreational value, exists than has for at least sixty years.'

On 28 October 1998 we again met Alan Meale. I attended the meeting together with Pat Leonard, Andrew Lyall, Vicki Elcoate and two of the Brighton MPs: Des Turner and David Lepper. The Minister said that neither he nor Michael Meacher had yet had time to study our papers, but they would soon be meeting to discuss the issues. Vicki Elcoate presented David Bangs' research paper on 'open country'. She also criticized the Countryside Commission for not carrying out a proper assessment of the South Downs, either by visiting or asking people who knew the area. Andrew Lyall presented the results of a National Opinion Poll survey commissioned by the Ramblers Association. This suggested that there was a huge level of public support for a South Downs National Park: In the area: 84% of the public were in favour and, nationally, of all those who knew about the debate, 83% were in favour. We also presented a catalogue of the worst damage through development and potential new threats from Eastbourne to Winchester. The Minister said he expected an announcement early in the New Year and promised to be *'as open as we possibly can be'*.

On 25 November Des Turner secured a debate in the House of Commons where he presented our case, brilliantly I thought, having clearly mastered all our key evidence. Unsurprisingly several local MPs from Sussex: Tim Loughton (East Worthing and Shoreham), Andrew Tyrie (Chichester) and Sir Geoffrey Johnson-Smith (Wealden) intervened in defence of the Countryside Commission's advice and the local authorities who were opposed to a South Downs National Park. Des Turner and David Lepper countered their arguments and emphasised the strong public support for a National Park.

In response, Alan Meale said that he had visited the South Downs. He had met many people who represented organisations and local authorities with an interest in the future of this outstanding landscape. He was particularly struck by how strongly people felt about the Downs, whether they were for or against a National Park. They all wanted a level of protection against major new developments that respected the importance of the area as a nationally recognised landscape and a resource for recreation and for biodiversity. The Countryside Commission had advised against a South Downs National Park and that decision reflected the Commission's assessment of the suitability of the area for national park status, which it had maintained since the 1950s. He gave an assurance that if the government concluded it was right to do so, it would ask the Commission to look again at the way it operated the criteria.

Our lobbying efforts continued: The Ramblers and Friends of the Earth organised a postcard campaign. David Lepper presented a 21,000-signature petition in support of the Campaign: no mean feat in the days when signatures had to be gathered manually!

At CNP two notable members of staff had been recruited: Ruth Chambers, with a chemistry degree from Oxford and a Masters from Manchester, where she studied environment assessment, planning and pollution, had a grasp of technical details that was to become invaluable to us. Her colleague Emma Loat, who had recently joined as a policy officer, was equally impressive.

On 15 March 1999 there was a very fruitful meeting with Michael Meacher. We were ably represented by Amanda Nobbs, Vicki Elcoate, Chris Todd and Ruth Chambers. Those that attended the meeting reported back that they were all appalled to find that the Minister appeared not to have seen any of our key documents. He did not appear either, to be aware of the level of public support for national park status. His officials tried to maintain that AONBs had the same protection as national parks in planning terms, something our representatives at the meeting hotly disputed. Michael Meacher himself was very perturbed by this (*'How am I meant to undertake a thorough analysis of the case'*) and told his special advisor so. He asked for a bundle of key documents to be sent to him at the House of Commons.

A few days later we wrote to thank Michael Meacher for inviting us to meet him and to emphasise what we regarded as some crucial points. We expressed our deep concern that the South Downs was being used as a political football by the Countryside Commission and the local authorities. Although we supported the Commission's desire to introduce legislation for the generality of AONBs, the South Downs Conservation Board should not be used as a stalking horse for this purpose. The Sussex Downs Conservation Board amounted to an extremely weak alternative. Their recent paper on planning stated that: '*The South Downs body should be in a position to take a single-minded view on the needs of the landscape of the South Downs and to pursue that untrammelled by the local planning authorities*' yet in the same paragraph they firmly asserted that '*the South Downs body should not be a planning authority and that the local planning authorities should remain responsible for structure and local plans and development control*'. Finally, we emphasised the scale and breadth of the support we had enjoyed in our nine years of existence.

At this same time, we heard that the Countryside Commission would be merged with the Rural Development Commission on 1 April 1999 to form a new Countryside Agency. We were delighted that one of the members of the new Board was Kate Ashbrook, Director of the Open Spaces Society, who was a staunch supporter of our Campaign.

Just when we thought that Ministers were perhaps becoming more sympathetic, we learnt that Lord Renton was going to promote a Private Members Bill in the House of Lords. Initially it was to be exclusively about the South Downs, but in its dying days the Countryside Commission had indicated that they would oppose it if that was the case as they were trying to get their Advice to Government turned into legislation. Lord Renton therefore introduced a Bill with four clauses from the Commissions' Advice regarding enhanced status for AONBs.

Lord Renton presented the second reading in the House of Lords on 21 May 1999. Inevitably, given the controversy surrounding the future of the South Downs, a number of speakers concentrated on that issue. Lord Morris of Castle Morris, Lord Beaumont of Whitley and Lord Addison all spoke in favour of a South Downs National Park, making use of information provided to them by CNP and CPRE. Some peers, resident in Sussex, praised Lord Renton and his Conservation Board. The remaining speakers spoke on the more general issue of the governance and management of AONBs. The views expressed ranged from those who felt that the majority of the nearly forty AONBs would not need purpose-built boards to manage them, to those who were against any further intrusions into the management of the countryside. Several speakers were concerned that the Bill did not sufficiently cover the issue of economic and social well-being in the AONBs.

I was particularly surprised to read the comments of Lord Kimball. I had

had a passing acquaintance with him when living in Lincolnshire and would not have expected him to take the stance he did in this debate. He said that 40% of the land was already scheduled as national parks, AONBs, SSSIs, green belt or National Trust land and that *'The awful truth is that countryside officers in various guises now outnumber farmers, farm workers and foresters. The offending officers do not understand the "living landscape approach" under which the countryside has developed'.*

At the end of the debate Lord Renton said:

'As regards the 77 per cent who voted in favour of a question put by the Ramblers Association in favour of national park status, I wish to put on record exactly what the question asked;

"If the South Downs was given national park status it would be run by an organisation whose duty would be to conserve the wildlife, natural beauty and cultural heritage, and to promote opportunities for public understanding and enjoyment of the South Downs' special qualities. Do you, yourself, think the South Downs should be given national park status or not?"

'How can anyone say no to that question? I am amazed that even 7% said no. If ever there was a loaded question, it is one. It is exactly the same as being asked whether one has stopped beating one's wife. I do not think it was a fair or particularly important point for the noble Viscount, Lord Addison, to make.'

Set out starkly in this manner, it is hard to challenge Lord Renton's assertion. However, it missed the point that the results of the Ramblers Association question reflected other evidence we had gathered of strong public support. Later public support was confirmed in the formal consultation during the designation process.

Lord Renton's Private Members Bill entered the House of Commons on 25 July, which was the last day for Private Members' Bills in that session. To our relief it fell, along with the rest.

It was becoming apparent that by this time our campaign was firmly on the political radar. The highlight of the Campaign for Vicki Elcoate was when she received an invite to No 10 to provide a briefing to one of Tony Blair's special advisers. She wrote that:

'I duly went along feeling quite anxious at the check in point at No 10, knowing that I probably had at best 15 minutes to sell the lengthy campaign and reassure the Government about the risks. The special adviser sat me on a sofa and asked me why we needed a South Downs National Park. It felt like a selling job, keeping it really high level and straightforward. And a few minutes later I was back on the street, not knowing if it had worked.'

This was followed by Vicki having a positive meeting with Michael Meacher who seemed supportive. However, he said that Richard Caborn, the Minister responsible for planning in the same Department, was not sympathetic, as he believed it would cause complications in the region's planning system.

Planning continued to be a major concern for us. We were worried on two counts. First was the continuing issue of the large number of planning applications being granted without, we believed, the protected status of the area being as fully considered as it should be because of the limited and non-statutory role of the Conservation Board: 'death by a thousand cuts'. Secondly, we were concerned by the number of major developments that could impact adversely on our two AONBs that were gaining traction, from road building to the Brighton & Hove Albion FC football stadium, from park and ride schemes to the Southern Water Portobello Treatment plant, to name but a few. At this time, though there was a presumption against such developments in a national park, no such safeguard existed in national planning policy for AONBs. Indeed, it wasn't until a Ministerial statement was made on 13 June 2000 that this safeguard was extended. An illustration of the issue, and our concerns over the Conservation Board, was the Sompting Waste Management Complex.

This was a proposal by West Sussex County Council to grant itself planning permission for a waste processing and transfer station within the AONB even though there was a site available outside the protected area. The Conservation Board's planning committee had recommended that the Board should object strongly and request that the Secretary of State should call it in for his determination. The Board members at the following crucial meeting were deeply divided and a motion was moved to defer the Board's decision, which would have meant it would be taking its decision after the County Council had already determined the application. Lord Renton, who chaired the meeting, voted for this amendment to the fury of a number of the members of the Board. This led to Chris Todd to call for his resignation and widespread criticism in the press. Annoying though this event was, it presented a golden opportunity for us to write a letter to Michael Meacher to illustrate the apparent lack of independence and weakness of the Conservation Board, whose planning powers were inadequate.

Though progress had been made, we were still far from our goal and depressed about the future of the Downs under the auspices of the Conservation Board.

CHAPTER EIGHT

THE BREAKTHROUGH

At the Labour Party Conference on 29 September 1999 the Deputy Prime Minister and Secretary of State for the Environment, Transport and the Regions, John Prescott, had announced that he was asking the Countryside Agency '*to consider designating in consultation with local authorities a National Park for the South Downs.*' He added that this, and a National Park in the New Forest, was '*a hundredth birthday present from Labour to the nation.*' We were later told that his civil servants had urged John Prescott not to make this latter statement because the designation process was in the hands of the Countryside Agency alone. Inevitably it raised the political temperature to boiling point locally and for many years ahead it would haunt us. But, at that time, it was a matter of celebration for us all.

Vicki Elcoate recalled that '*Our little band had headed off to the Association of National Parks' conference in Pembrokeshire, leaving Emma Loat in the office. Chris Bonington, Ruth Chambers and I were on a walking tour, admiring the beautiful coastal scenery, when we got a misspelt message from a clearly highly excited Emma. John Prescott had announced the intention of the Government that the South Downs should become a national park. At the moment we were sharing the message around a rainbow appeared over the cliffs – marking the perfect climax of the most incredible campaign carried out locally, nationally and involving every kind of tactic you could think up*'.

More widely, reaction to the announcement was decidedly mixed. We were encouraged by that of the Hampshire Wildlife Trust and the National Trust. The Hampshire Wildlife Trust had originally opposed a South Downs National Park, and though it now changed its stance to one of support, this never seemed more than lukewarm, probably because it didn't want anything to get in the way of the creation of a New Forest National Park. The National Trust wrote us a letter of support, which was very welcome given that its attitude had been ambivalent up to this time.

Others greeted the news less graciously. The Country Landowners

Association declared that local communities, farmers and landowners were alarmed by reports that the Government was about to announce the designation of the South Downs as a national park: *'Giving national park status to the South Downs will not give one iota more protection than it already receives through its existing status as an AONB'.*

Peter Bryant, Chairman of the Sussex Rural Community Council and a member of the Conservation Board said: *'The Downs should not be treated as an isolated island with a separate planning designation. A national park would be an arbitrary imposition and is not wanted by the majority of people who live in the Downs.'*

Once more we had to write to the local papers to correct a press statement from West Sussex County Council. We had to refute the contention that a national park authority was an unelected quango with little experience of local planning and a chairman appointed by central government.

By contrast, and somewhat to our surprise, a press release from the Sussex Downs Conservation Board said that the Board was glad that the Government had finally made a decision on the future management of the South Downs:

'It has been widely recognised that the Conservation Board has successfully defended and enhanced the Downs during the past eight years. We now hope that there will be a smooth transition between the Conservation Board and the National Park Authority to ensure the landscape is conserved for future generations.'

Despite this, we were very conscious of the continued opposition to the National Park from a number of the Conservation Board members, so we awaited the next meeting of the Board in October with some interest. Len Clark wrote afterwards that:

'Most people were expecting something approaching an indignation session, since the majority of the local authorities (who comprise two thirds of the membership) had rejected the national park option. However, to the surprise of some of us not a word was said in protest. The way had been apparently smoothed by the conciliatory response of Chris Mullin (who had taken over from Alan Meale) to a deputation from local authorities to him on 12 October. It is likely that the question of delegation of planning powers will be prominent in future negotiations.'

On the same day of John Prescott's announcement, Michael Meacher wrote to the Countryside Agency. I thought it a masterly document. Certainly, it had huge significance, so I reproduce here the main points with regard to the South Downs:

'Turning to the South Downs, the Countryside Commission advised that this area does not meet the criteria for designation as a national park, as presently defined and applied....I would like to request the Countryside Agency to review this position, in particular how the Countryside Commission has applied the 1949 Act criteria (though not the criteria themselves).

In designating national parks, I understand the National Parks Commission previously placed emphasis on selecting areas of open country with a degree of ruggedness or wilderness. This was appropriate 50 years ago, when few urban dwellers could visit the countryside regularly and the range of recreational provision was very limited. An emphasis on rugged and open country is less appropriate today, when millions go in search of fresh air and a wide range of recreational pursuits in less remote countryside. The Government does not wish to discourage visitors to the countryside: indeed we are committed to improving people's enjoyment of it.

As those who framed the legislation recognised, some areas will require special management by a national park authority, to conserve and enhance their natural beauty, wildlife and cultural heritage, and to promote opportunities for public understanding and enjoyment. In the Government's view, it is no longer appropriate to limit such management to predominantly open and rugged countryside. Further, we consider more account should be taken than in the past of the need to provide for improved opportunities for open air recreation for the population at large, including by providing recreational opportunities where people live.

I note that, amongst the areas originally recommended for designation as national parks by Sir Arthur Hobhouse in 1947, the South Downs is the only one not to have been designated as a national park (or equivalent, in the case of the Broads).

In the light of the policy outlined above, I ask the Countryside Agency to reconsider its current policy with regards the 1949 Act criteria, and to look again at the case for a national park in the South Downs. It will be important to consult local authorities and other interests as to how practical arrangements might best be made. In assessing the priority to be given to this, the widespread support for a South Downs National Park, which has persisted for many years, should be considered.

I would also like to make it clear that I have every intention of proceeding also with a package of measures to enable better management of AONBs, drawing on the Countryside Commission's advice 'Protecting our Finest Countryside'.

I wrote a personal letter to Michael Meacher to thank him for what I believed to be a courageous decision, given the considerable opposition he had met on several fronts. I assured him that we would do everything we could to ensure that people had a clear understanding of the issues under debate.

All eyes were now on the Countryside Agency as we awaited its response to the Minister's letter.

We were therefore extremely concerned by a paper prepared for the Agency's Board meeting on 16 December 1999. This set out officials' first thoughts on the South Downs and an outline of the processes to be followed for future national park designations. It suggested that, in assessing the potential of new national parks, precise measurements should be used rather than the more subjective approach of the past. For instance, a maximum of 30% of lesser quality land is acceptable in a national park. A minimum of 25% open access land, excluding woodland, is needed to meet the recreational objective. Access should be measured by a maximum of 1.5 hours by public transport from a conurbation greater than one million population.

We only became aware of this paper at the last minute. Alarmed, Vicki Elcoate hastily arranged for a letter and paper to be sent to Ewen Cameron, the Agency's chairman. These were pushed under his door at 11.30pm on the eve of the Agency's meeting. I also sent him a letter by fax from the Campaign. We urged the Agency's Board members not to endorse the current recommendations of their officers, to press ahead with the designation process and to consider the case for a South Downs National Park at the earliest opportunity.

At the Board meeting Frances Rowe (Board member with responsibility for protected areas), welcomed the opportunity to review protected areas for designation. She said that there was an exciting opportunity to help individuals to reconnect with the landscape. The Bishop of Blackburn, another Board member, said that it would not be easy for the Agency to come to a different conclusion to that of their predecessors, but an announcement by Michael Meacher made at the Labour Party Conference about the restoration of downland could help to get them off the hook.

Kate Ashbrook and some other Board members were very critical of the paper in front of them – it was too prescriptive. There was no mention of Hobhouse's important point about the spread of areas. Woodland was not mentioned as a source of open access recreation. On the other hand, members and officers were understandably concerned that, if they were not careful, a number of AONBs might fit the revised criteria used for assessing potential national parks. The paper was sent away for revision.

Our own members shared the criticisms of the quality of this paper. We had heard that various experts including Professor Ron Edwards (of the Edwards Review) had been consulted by the Agency about the paper's content. We also heard that there had been dissention amongst the Agency's officers about the content of the report, though we had no way of checking this out.

We were also unhappy about the lack of consultation or dialogue that the Countryside Agency was having with stakeholders other than the local authorities. In particular, we were concerned to learn that they were establishing a working group to look at planning in new national parks with members from central government, local authorities and the existing national park authorities. No one from the voluntary sector, not even a highly visible campaign like our own, had been formally told about the existence of the group or had been invited to be on it.

Accordingly, in early January 2000, we wrote to Ewen Cameron, the Countryside Agency chairman, saying that the paper that had been presented to his Board would, in effect, disqualify some existing national parks, let alone provide adequate provision for the creation of new ones. We also expressed our great concern that the Countryside Agency had not liaised with non-governmental organisations for many months despite a continuing dialogue being undertaken with the local authorities.

Pat Leonard and I had a constructive meeting with Victoria Edwards, a Countryside Agency Board member, the day before the Board considered a revised paper on national park criteria, which we also found wanting. On 10 February 2000, the Board met and agreed that the revised paper could go forward, subject to some amendments. In it were some principles that would determine future national park designations. It was suggested that in areas already designated as AONBs the natural beauty was likely to be a given. Therefore, the key focus of the designation process should turn on two key questions:

a) Is the area concerned an extensive tract of country providing or capable of providing sufficient opportunities for open air recreation? The area clearly needs to have characteristics that mark it out as different from the bulk of "normal countryside"; so it needs to be more than a simple network of rights of way. It should contain qualities that might merit the provision of a markedly superior recreational experience.

b) Is it especially desirable to provide for the leadership of a national park authority, with the powers and duties laid down in the Environment Act 1995? In other words, will designation lead to a markedly better managed recreational experience than can be achieved by local authorities alone. And will this recreational experience be available, promoted and interpreted to the "socially excluded" as well as the more mobile in society, as a result of the work of that special authority?

This paper did not address the specific issue of a national park in the South Downs. That was to be considered at a further meeting in April.

Shortly after this meeting we had another one with Michael Meacher. We congratulated him on the Government's announcement on the South Downs. In reply he said that he thought ours had been a *'splendid campaign'* and that it was *'a model of the way in which things should be done'*. He said that the Campaign had played *'a very real part'* and that we should take credit for what had been decided.

Having assured the Minister of the Campaign's support we went on to express our unease about the timescale being proposed by the Countryside Agency. It was emphasised to us by officials present that the Ministers only came in at the end of the designation process and before then there would almost certainly be a Public Inquiry to examine the validity of national park status, its boundary and its administrative arrangements. Such an Inquiry would be for the Department to manage, not the Countryside Agency. Prophetically it was said: *'Most Public Inquiries do take a while!'*

Following the meeting Michael Meacher wrote to me with reassurances that we would have the opportunity to feed in our views on planning. He wrote:

'We have recognised that the context for planning control in a South Downs National Park is complicated by the shape of the area being considered and the number of local authorities potentially involved. This is why we have given an undertaking to look carefully to see whether any departure from the standard model might be justified.' He ended by saying that *'there would be continuing opportunities for you to feed in your views and I am sure you will do so.'*

The paper before the Countryside Agency Board for its meeting in April 2000 proposed that the process of designating a national park in the South Downs should now begin. The paper concluded that there was sufficient land in the broad area of the two current AONBs that met the criteria. It recommended that work should commence to identify a draft boundary and hold a consultation on this and other aspects of the designation. They would then receive a further report before deciding whether or not to make a Designation Order. Meanwhile the Board should continue to support the work of the Sussex Downs Conservation Board and East Hampshire AONB Joint Advisory Committee. There was also a paper with proposals for downland restoration.

The Countryside Agency's Board members were to visit the two AONBs the day before the meeting. John Templeton had written a long letter in late March suggesting places they might visit. He got a response thanking him for this but observing that the day had been cold, dark and wet! Nevertheless, they had been impressed by what they saw despite the poor weather.

The Board meeting was held in Winchester on a chilly day. I can recall sitting in the public gallery with Len Clark, John Templeton, Owen Plunkett (from the Hampshire Ramblers' Association), Emma Loat, Emily Richmond (from the Ramblers Association nationally) and others to listen to the debate. The Board members professed to see the potential of the area as a national park, both in terms of recreation and the enhancement of its special qualities. Victoria Wood and Kate Ashbrook spoke in favour. Frances Rowe, with experience at the Northumberland National Park, could see the great benefits that a national park could bring, particularly in the South-East of England. The Board members were conscious of the conflicting views of local people and emphasised that the process must not be rushed; and the consultation process should be thorough. Ewen Cameron described the landscape as outstanding, the element of doubt was whether it was 'especially desirable' to have the leadership of a national park authority. To our great joy the Board unanimously decided that the South Downs met the criteria for designation as a national park and that the process of designation should proceed.

In our press statement welcoming the Agency's decision we emphasised that our main concern was over the question of planning and the importance of a South Downs National Park Authority having full planning powers, that would enable it to balance the national importance of the area with local needs and ensure that planning was delivered in an accessible and locally accountable way.

After all our long struggles with the Countryside Commission, it was extraordinary to read the *volte-face* in the Countryside Agency's press statement:

'When the first national parks were set up fifty years ago, they were selected partly for their wild, rugged qualities. Times have moved on, and we now have a tremendous opportunity to create a national park for the 21st Century, one which meets the demands of a much more mobile and leisure conscious population.

The Downs offer an exceptional recreational resource and the chance to get away from it all. They are accessible (often by public transport) to a large urban population. National park designation would bring a National Park Authority with the resources and powers to help visitors enjoy the area sustainably. It would also enable farmers, landowners and local people to benefit from the opportunities offered by national park status.'

It was spelt out that:

'The first step will be to develop further the vision of what a national park can achieve, to help public understanding and identify areas which might benefit most within the two Areas of Outstanding Natural Beauty. Then we will proceed to define a draft park boundary for a National Park for the South Downs. We will also consider carefully the best tools for a future National Park Authority to maximise the potential of this special area. We need to discuss how planning should be handled.'

Whilst the Countryside Agency moved forwards the designation process, we considered the future of the Campaign itself.

I had drawn up a seventeen-page discussion paper on its future for our meeting in February 2000. Its purpose was to assess the needs and priorities of the group and its supporting organisations in sustaining an effective campaign and achieving its goal. I included a robust analysis of our strengths and weaknesses and the potential threats we might face in the future. Our main weaknesses were our lack of influence with local authorities and the Countryside Agency and our lack of a meaningful presence in Hampshire. The potential threats were many and varied: some, like a move by local authorities to minimise the size of any national park became all too real: others, like a potential change in legislation that would jeopardise the chances of a South Downs National Park, were thankfully more speculative.

Before that meeting took place, Vicki Elcoate facilitated a very constructive meeting of nationally based organisations with the purpose of planning a successful campaign for the New Forest and the South Downs over the following three years. CPRE, the Youth Hostels Association, Friends of the Earth, Ramblers Association, Open Spaces Society and the National Trust were all represented. The most important decision taken that day to was to agree two key points of principle which we hoped would be adopted by all our members. The first was that the proposed South Downs National Park should include the whole of the existing Sussex Downs and East Hampshire AONBs. The second was that the new National Park Authority should be vested with full planning powers. We were gearing up for what was to follow.

CHAPTER NINE

FIRST MOVES TOWARDS DESIGNATION

We held our first conference in the spring of 2000 at much the same time that the Countryside Agency circulated their publication 'South Downs News'. In it, they set out the timetable for the designation process:

May to August 2000	Agency discuss and agree their vision
May 2000 to January 2001	Agency assessment of area that would most benefit from national park status and stakeholder engagement
Sept 2000 to Sept 2001	Working Groups, seminars etc. to consider how a national park would be run including planning
March 2001 to August 2001	Identification of Draft Boundary
September 2001 to December 2001	Public consultation on Draft Boundary
September 2001 to December 2001	Discussion of options for administrative arrangements
February 2002 to April 2002	Consultation with Local Authorities on Draft Boundary
February 2002 to April 2002	Consultation on administrative arrangements
Spring 2002	Agency makes Designation Order and submits it to Secretary of State

We were concerned that this timetable seemed protracted, and a lot of time devoted to developing a vision. We thought one possible explanation was that the Agency saw it as a way of solving any disagreements through rational argument at the beginning of the process, which would then speed up the procedure further down the line. On the other hand, it could be that the Agency was determined to do the job thoroughly to avoid judicial review. A more sceptical explanation might be that legislation in the Countryside and Rights of Way Bill (later known as the CROW Act) on AONBs could be presented as an alternative to the National Park. After all the setbacks our Campaign had suffered in the past it was perfectly understandable that we were extremely anxious about the future of our hard-fought crusade. But we remained determined and to further our goal by engagement with the Countryside Agency, Government more generally and with the public.

Our first task was to make progress in Hampshire. As the designation process got underway, we renewed our efforts to engage with voluntary groups there. This brought back into play yet again the issue of the size of a national park and whether it should be based on the two AONBs, as intimated by the Countryside Agency, or be restricted to the chalk ridge as favoured by the local authorities. Up until now our efforts had been very much concentrated on East and West Sussex. In East Sussex the issue of a 'chalk-only' national park was academic, since the AONB hardly extended into the Weald in that county. In West Sussex it was a major issue, with the market towns of Midhurst and Petworth and the bulk of the population in the AONB being situated in the Western Weald. Hampshire was different again. Virtually the whole of the East Hampshire AONB was on chalk and sparsely populated, but the western extremity of the Western Weald lay in that AONB. This contained Petersfield, the largest town in any AONB in England or Wales, and Liss, the largest village in either of our two AONBs.

The position of Hampshire County Council had been much less antagonistic to the concept of a national park than West Sussex County Council, although issues had been raised around planning. Later they confirmed their support for a South Downs National Park and played an important part in the final stages of the process.

Only two district councils had land within the East Hampshire AONB, East Hampshire and Winchester. In both cases much of the land area of the districts lay within the AONB and so national park status threatened to impact them more than any other district with the possible exceptions of Chichester and Lewes. Consequently, both districts were opposed to the notion of a national park at this stage. The 'chalk-only' argument did not impact Winchester, but it was a big issue for East Hampshire given the population density in its part of the Western Weald. It therefore sided with the notion that if there was to be a South Downs National Park it should be confined to the chalk ridge, thus effectively protecting its planning role. It had responded to John Prescott's announcement thus:

'...*The creation of a South Downs Planning Authority would just add another layer of bureaucracy to everyone's lives. Just imagine the difficulties for local people if day-to-day planning decisions for about half our District area were made 50 miles away in Lewes, Sussex – or even further afield.'*

No doubt to reassure the residents of Petersfield and Liss, a paper presented to its full council suggested that the proposal for a *'chalk only'* national park would create *'a 'recreation-based' national park and a 'conservation-based' AONB'*.

Not all the councillors bought into this narrative. Minette Palmer, a Conservative member from the iconic village of Selborne, championed the National Park proposal and challenged the received wisdom. Such was the passion generated by the issue that it led to her resignation from the council. Their loss was our gain because she later joined the Campaign and was actively involved on the Executive.

Against this background we ventured to build our support in Hampshire. We were fortunate to have help to hand from the Hampshire Ramblers. The chair of their South-East Group, and later chair of the County Group, Owen Plunkett, had been a member of our Campaign committee for several years. He had joined the Ramblers because of his passion for the countryside and walking in it and was a member of the Ramblers' Board of Trustees. Having retired from his job as Senior Lecturer in organic chemistry at the University of Portsmouth, he had both boundless energy and the necessary time to lead 40 'South Downs Walks' and give over 30 talks to help promote the cause. He had been keen since he joined us to see the Campaign become more involved in Hampshire, thinking quite rightly, that we were too Sussex focussed.

But we needed to expand our influence further. In total contrast to the Sussex Wildlife Trust, the Hampshire Wildlife Trust continued its seemingly half-hearted support and maintained that planning control should remain with the county, which was not in accord with the national Wildlife Trusts partnership statement on the South Downs. We therefore had to look elsewhere.

Our initial approach to CPRE was not auspicious, since the Hampshire branch was concerned that the Campaign seemed to them to be too Brighton focussed. But in June of 2000, Stephen Harwood, the chair of the East Hampshire branch, organised a meeting for their members at the Queen Elizabeth Country Park, situated to the south of Petersfield adjoining the A3 at Butser Hill. Planning Officers and Councillors from Winchester and East Hants also attended. Pat Leonard, Amanda Nobbs and I were invited to speak on behalf of the Campaign. We described the structure, management and purposes of a national park and the benefits it could bring to the area currently designated as an AONB. We also outlined the proposed programme for designating a South Downs National Park.

Successful public meetings were held in Winchester and Petersfield. We then had an invitation to address the East Hampshire Association of Town and Parish Councils. Amanda Nobbs and I both spoke at their meeting on 7 September at Greatham, a village to the north of Petersfield. Their secretary sent us a warm thank-you letter in which he informed us that there had been a lot of opposition to the national park proposal, and only a small number of councils were supportive. He added, however, that since only 19 of the 40 parishes and town councils were present, it might show that 21 local councils were not sufficiently worried to get any of their members or clerks to attend. One unsolicited compliment came from a member of the audience who said that it was the first meeting at which he had heard everything that the speakers said!

Later Stephen Harwood sent us a memo to emphasise that, in East Hampshire, there was a strong feeling among parishes, as well as the District Council, that the Local Plan must stay with the District. This view was also reflected in their local membership and more generally. We should not be under the misapprehension that it was 'just Local Councils wanting to keep their planning empires'. Fortunately, this negativity changed radically over the next eighteen months.

As we engaged with communities in Hampshire, so too did the Countryside Agency with stakeholders across the whole area. In July 2000 we were invited to attend a Countryside Agency meeting at the University of Sussex 'for user groups and conservation bodies' where we were briefed on the main topics on which the Agency would be consulting.

This meeting was one of a series held by the Countryside Agency over this summer period. Following a meeting with the local authorities in June, West Sussex County Council, whose opposition was unrelenting, wrote on behalf of twelve local authorities to give their views on how the planning function might best be organised. Whilst acknowledging that national park authorities include elected members from all tiers of local government, they stated that the direct relationship between the electorate and their representatives would be lost as there might be only one member from each district council area. Many of the affected councils had brought development control closer to their residents through local area committees. Because of the linear shape of the proposed national park there was a risk of remoteness from a national park headquarters and regional offices would add to the costs. Whilst the cost of providing full planning services by a national park authority would not fall directly on local tax-payers, it was estimated that the additional cost would be approximately £1M per annum and the addition of a national park authority would add an extra tier of government.

The local authorities recommended that there should be three separate structure plans, one for each county, prepared jointly by the existing strategic planning authorities, the counties, and a national park authority. They asserted that minerals and waste planning should remain with the existing authorities, again the counties. For the Local Plan, which was the domain of the district and borough councils, they recommended joint planning arrangements in each county area, to provide three plan making authorities.

We had anticipated that there would be big push-back on planning from the local authorities and so, just four days after the historic decision had been taken to commence the designation process, we had met and decided to set up a planning group under the chairmanship of Ruth Chambers to consider how we would address planning issues in the succeeding phases of that process. We also decided we needed to do some work on the boundary, both to try and influence the Countryside Agency's criteria and the scope of a national park. We were pleased the Countryside Agency seemed to be taking the two AONBs as the starting point, but felt we needed to support this and consider what areas outside the AONBs might merit inclusion. Emma Loat, John Templeton and Paul Millmore offered to form a boundary working group with someone from Hampshire. Later Terry Smith of CPRE Hampshire joined them.

They took to their task with great alacrity and brought back to our Executive ambitious proposals. Paul Millmore wanted the boundary to be extended into the sea from Beachy Head to Brighton and up the river valleys and to include transport hubs. John Templeton wanted to include a sizeable chunk of south Hampshire based around the Royal Hunting Forest of Bere as well as parts of the Surrey Hills AONB. The Executive decided that they should prepare an outline case for the inclusion of different broad character types based on the Countryside Agency's interpretation of the boundary setting criteria, and that they should concentrate their efforts on Hampshire and the Weald. John Templeton carried out a desk study of the whole area whilst sitting in his garden with, as he recalled, rosy-red apples falling around him. When he brought back his proposals, we accepted almost all of them and a rather less ambitious list of 30 possible additions outside the AONBs was agreed. John Templeton and Emma Loat worked frantically to get our proposals to the Countryside Agency's consultants by their September deadline for consideration.

In that same month of September 2000, our representatives met with officers from the Countryside Agency. The Agency suggested that since they had embarked on the process of designation there was a 90% certainty that the National Park would be confirmed and so there was no need for us to go on campaigning for the principle. They believed that the Public Inquiry would largely be about the boundary though it was likely to stray into issues like the impact on planning. Public consultation on the boundary would take as a starting point the two AONBs and would look at areas beyond that which might be considered for inclusion. Potential for enhancement would not be a sufficient reason for including areas outside the AONBs . If parts of an AONB were left out they would have to be of a viable size to operate as a stand-alone AONB or be able to become part of another AONB to avoid de-designation. As an aside we were delighted to learn that the local authorities had complained at their meeting with the Agency that they were having to fight against a well-resourced and effective campaign by the South Downs Campaign!

By October 2000 we had made tentative steps into Hampshire, had discussions with the Countryside Agency, set up a group to progress our ideas on planning to counter those of the local authorities, and provided the Countryside Agency with our views on which areas outside the AONB boundaries should be considered for inclusion in a national park. But another issue, which had been bubbling since the early summer, now came to the fore. It arose because the decision on how to organise planning once a national park authority was set up rested, in law, with that authority. The local authorities were clearly nervous that even if they succeeded in persuading the Countryside Agency and a Public Inquiry of the merits of some form of delegation or joint working a national park authority could still decide to undertake all planning itself. Led by Hampshire County Council, there was a move to use the Countryside and Rights of Way Bill to remove this possibility by, in effect, transferring that legal power to central government.

In the summer of that year, we had been very encouraged when we saw a letter sent from the Countryside Agency's chairman to Chris Mullin following a discussion at the Agency's May 2000 Board Meeting. In it he said that the Board was opposed to any dilution of a national park authority's responsibilities for the local plan preparation or development control and therefore its advice was to resist any relevant amendments put forward by Hampshire County Council to the Countryside and Rights of Way Bill. He went on to suggest that there were options for delivery of these responsibilities, including delegation, which could be considered as part of the detailed administrative arrangements.

This did not deter the opposition. During a debate on the Countryside and Rights of Way Bill on 16 October in the House of Lords, the Earl of Carnarvon stated that *'the Secretary of State can make an order under the Section 67(2) of the 1995 Environment Act that land-use planning should remain with the existing planning authorities. Where he does not, he encourages the National Park Authority to make voluntary arrangements to work with the neighbouring planning authorities to prepare a joint structure plan for their combined area'.*

'There is no provision in the Environment Act for the Secretary of State to make an order for development control functions - which I stress - to remain with the existing local planning authorities in areas when there is a national park. The local authorities in the New Forest and South Downs areas have been examining the implications of national park designation, including the implications of existing planning functions. The authorities there are proud of their activities in protecting the New Forest and the South Downs from inappropriate development. Amendment 542 would enable the Secretary of State to make an order that development control functions should remain with existing planning authorities if considered appropriate in the circumstances of the areas.'

This amendment was supported by the Earl of Selborne and Lord Renton who said *that 'It was a lacuna in the Secretary of State's powers in that national parks have the ability to delegate development control to local authorities in their area if they wish to do so; but the Secretary of State does not possess that power.'*

In a letter written by Ken Thornber, Leader of Hampshire County Council, to the chairman of the Countryside Agency, he stated that the Government Minister recognised the intention behind the amendment. Lord Whitty had explained that rather than legislating generally in the Bill now, the Government *'is intent to wait and see what is required in these two specific situations (The New Forest and the South Downs). The Government did not have a closed mind'*. The amendments were withdrawn. Hampshire County Council sought assurances from the Countryside Agency that it would not rule out solutions which required legislative change in connection with the administration of planning in either the New Forest or the South Downs.

This was followed in November 2000 by a debate in the House of Commons in which David Lepper presented our case for a South Downs National Park. Nigel Waterson, MP for Eastbourne advised the government and others involved *'to be cautious about divining public opinion or opposition. There are some vociferous minority groups who are in favour of the proposal, and there have been some so-called public meetings that have not been representative of a cross-section of opinion. In my constituency, mainstream opinion is not in favour of a national park.'*

David Lepper recalled a number of Conservative MPs expressing concern about the loss of planning control by local councils, extra bureaucracy and so on and pledged the designation process would be terminated if they won the General Election, expected the following year. Fortunately, the opponents of the National Park and of the CROW Act did not have an opportunity to prevent either happening and the designation process continued unabated.

CHAPTER TEN

NEXT STEPS

In 2001 the designation process increased in momentum. It began with a two-day seminar in Chichester in January attended by many public, private and voluntary bodies with an interest in a national park.

The boundary setting process was explained by Professor Robert Tregay, Senior Partner of Landscape Design Associates, whose team was contracted by the Countryside Agency to carry out the work. Little did we then appreciate what an important figure he was to become for us later in the process. He explained that his team had first identified a broad area that they believed met the Countryside Agency's criteria, known as the Area of Search. A draft boundary would then be identified within this area. They approached the work in a very systematic way based on legal criteria. He said that they had concluded that the two AONBs met the natural beauty criterion for inclusion in the Area of Search, and to our delight added that most of the areas we had proposed merited consideration at this stage. The next stage, of identifying the draft boundary for consultation, would follow once the Area of Search had been approved by the Countryside Agency.

The Countryside Agency had, by this time, set up a number of Technical Advisory Groups to consider issues related to the designation process. We were glad to be included in this work.

A Technical Group on Planning was dominated by the local authorities. This group's task was to consider three options. The first, which would require primary legislation, was to maintain the status quo with all planning carried out by the local authorities. The second was to have joint structure plans, one for each county, the county councils being responsible for all minerals and waste planning with development management resting with the National Park Authority. The final option was for the National Park Authority to have responsibility for all planning.

Ruth Chambers was invited to provide evidence to this group and she produced what I thought to be an excellent paper that pointed out that if the National Park Authority had full responsibility for planning it could enter into arrangements with existing planning authorities, citing cases where this had happened in the Dartmoor and Northumberland National Parks.

Ruth and I were invited to join the Countryside Agency's Governance and Administration Group. Despite its membership being largely from local authorities there was a remarkable degree of consensus. We agreed that the National Park Authority needed to provide strategic direction, but also should be sensitive to local variation and local needs. It needed to produce a strategic vision that engaged ownership and delivery by others. Whatever the final governance structure, it was agreed that, because of the geography and scale of the National Park, there might well be the need for a significant degree of delegation, or contracting out, of duties to existing authorities. We were also agreed that there was no justification for moving away from the existing model for national park authorities and if any improvements were thought necessary, these could be fed into the current 'Review of National Park Authorities'. There was a majority in favour of the authority choosing its own chair and having a committee structure supported by topic groups with co-opted members from the local community.

Although it had become abundantly clear to us a year earlier that we would need to up our game if we were to play a full part in the designation process, and that we must employ someone to help to deliver the Campaign, by this stage we still did not have a Campaign officer and the strain was beginning to tell. The recruitment process had taken longer than we had anticipated: the stumbling block had been a financial one. We had applied to a number of our member organisations and charitable trusts, and it took time for them all to respond. But by January 2001, I was able to report that we now had sufficient funds to enable us to recruit a Campaign officer. We had received a grant of £10k per annum for two years from the Kleinwort Trust, a promise of £5k per annum for two years from the Sussex Wildlife Trust and the Society of Sussex Downsmen and £3k per annum from CNP for the period of the post. We also received significant funds from CPRE nationally and locally. It was very welcome to see the donation from the Downsmen, who had re-joined the Campaign now that a designation process was underway.

When we met again in April 2001, I was delighted to announce that five good candidates had been interviewed and that Chris Todd had been appointed as Campaign Officer. He would be based at his home and would be working twenty hours a week: that, at least, was what his contract said. In fact, he worked many more hours than that. Chris had already been very active in the Campaign. A tall man with a shock of red hair, he always practised what he preached and frequently put some of us to shame as we continued to eat meat and drive cars.

He had an honours degree in aeronautical engineering and a post graduate diploma in Environment Impact Assessment. We were extraordinarily fortunate that we were able to appoint a multi-talented person who already had a deep knowledge of, and passion for, the South Downs. He was not only an extremely efficient administrator. His assiduous attention to detail helped to ensure that our papers were always thoroughly researched. He was also an astute organiser who dramatically increased our membership over the years ahead. All of this was achieved whilst he, very slowly, virtually rebuilt his house in Brighton. Members of the Campaign returned from visits there reporting that the staircase had disappeared or there were seemingly no floorboards left on the ground floor. The housebuilding took longer to complete than the successful Campaign but was equally painstakingly well-managed.

There was no time to spare. Just before Chris Todd began his work for us in April 2001, the Countryside Agency announced their decision on the Area of Search.

We waited eagerly to see what had been decided and viewed the results with decidedly mixed feelings. Most of the two AONBs were included, though Petersfield and parts of the 'A3 corridor' to the north of the town were proposed for exclusion. Some further minor variations to the existing AONB boundaries were also proposed. Some broad areas outside the AONBs around the North Itchen Valley, the Hampshire Hangers and the West Sussex Low Weald were included. Although we were pleased to see the proposed additional areas, we were disappointed that not all our recommendations had been accepted, most notably the whole of the Heritage Coast between Seaford and Eastbourne, the lower Arun valley, the Climping Gap and the area north of the A27 between Slindon and Chichester.

But it was the deeply worrying exclusion of the part of the East Hampshire AONB including Petersfield and the so-called 'A3 corridor' to the north that led to our breakthrough in Hampshire. The area in question contained not only Petersfield but also Liss, a major road, the A3, and the Portsmouth to London railway line. It had been included within the Area of Search by the consultants but was mysteriously omitted by the Agency itself.

Despite the area being in an AONB, it strong transport links meant it was under considerable development pressure: for example, previously there had been calls to amend the Hampshire structure plan to allow 1,500 houses to be built at Liss. Voluntary groups in Petersfield and Liss were therefore horrified at the prospect that their settlements could lose all protection and vowed to fight the exclusion.

This potential exclusion brought to the fore Margaret Paren who was to become an outstanding leader in the Campaign. She was a retired Civil Servant who had spent most of her career in the Ministry of Defence. As many before and since have noted, she made up for her lack of stature by her determination and strength of character. She recalled for me the start of her involvement in the South Downs when her very tall retired Civil Servant husband Nigel Paren returned from what should have been a routine Parish Council meeting:

'I heard the back door slam and Nigel appeared shortly afterwards carrying two large glasses of red wine. He put one into my hand and said: "We have the biggest fight ever on our hands and only a 10% chance of success" He then went on to tell me that Liss had been omitted from the Area of Search. He took his glass of wine upstairs to the study and drafted a letter to the Countryside Agency on behalf of the Parish Council. By the following morning we had resolved to fight this outrage and begun to assemble a group of like-minded villagers. The rest is history.'

A group was thus formed in Liss. They discovered that the Countryside Agency would be holding a public meeting in Petersfield and decided to attend to make their case. They were enthusiastically supported in this by Dr Francis Rose who well remembered his earlier associations with the designation processes for the AONBs. By now he lived in Liss. He was a large, bearded man with a quiet voice and a monumental knowledge of plants: his favourites being the lower plants that interested few others, like the ferns, lichens and mosses. He wrote a number of highly regarded books, but his tremendous standing amongst environmentalists did not stop him engaging with his local community and he was very angry about what had happened. Unable to attend the meeting himself because of ill-health, he asked Margaret Paren to read out a passionate plea from him to the Agency to think again. After the meeting Owen Plunkett, who attended the meeting on our behalf, caught up with Margaret and said that they needed to talk. Margaret joined our next meeting and later became the Hampshire CPRE representative. After the unpromising start the Hampshire branch of CPRE became one of our most effective member groups.

We now had a major task on our hands as we needed to respond to the Area of Search proposals from the Countryside Agency. A team under John Templeton was convened and Chris Todd took on an ever-increasing role. In preparation for the forthcoming public consultation, they scrutinised the whole of the published Area of Search, taking into account as they did so the views of our member organisations. They had to ensure that any areas that they felt merited inclusion met the criteria set by the Countryside Agency. They then had to write a detailed analysis of each area.

John Templeton, in particular, became very heavily engaged in the process. He began his work in late April 2001, meeting with members of Hampshire CPRE about the exclusion of the A3 corridor. During May he continued his review of the areas supported by Hampshire CPRE before moving on in June to consider areas in Sussex. With the end of June deadline for comments to the Countryside Agency looming, he and Chris Todd undertook an epic six-hour drive through Rottingdean, Newhaven, Seaford, Alfriston, Lullington, Abbotts Wood, Milton Hide, Upper and Lower Dicker, Berwick Station, the Adur Valley, Steyning, Spithandle Lane, Henfield, Woods Mill, Windmill Hill, Woodmancote, Hassocks and the Jack and Jill windmills.

Chris Todd put together a draft response, over 50 pages long, and circulated it to John and the others on 27 June. John recalls having been up to 4.20am, sustained by cups of coffee and the loyal companionship of his Yorkshire terrier.

The boundary group's efforts were graphically described in an e-mail to me from John:

'Thank you for your kind words about the "team". It was one of the most exciting and fulfilling pieces of work I've ever been involved with, and I wouldn't have missed it for anything! I've found it immensely stimulating working with Chris and Paul (Millmore) from the start of the boundary exercise, as well as the two intensive days last week spent in Chris's house, emails were flying backwards and forwards between us until the last minute!'

Having submitted the massive document supporting our proposals for the 30 additional areas, Chris Todd's next task was to write a newsletter containing a double-page spread showing the two AONBs, the Area of Search and our proposals. This kept our member organisations in touch with what we were doing and provided them with contacts for different parts of the boundary that they could approach for more information as necessary. The newsletter also provided an opportunity for spreading the word more widely. In this regard, it was good to see a supportive feature in the West Sussex Gazette a week later.

The Countryside Agency Board agreed in September 2001 to hold a public consultation, and this was launched on 27 November. The consultation document was accompanied by a draft boundary. The consultation was to close on 28 February 2002, which sounded a reasonable length of time but, of course, included Christmas and the New Year. There was much to do. The draft boundary included some areas we had proposed, either wholly or in part. Most notably Petersfield and the A3 corridor had been included, though Chris Todd remembered Margaret Paren commenting, presciently, that she wasn't going to celebrate yet, there was a long way to go. We were also glad that, despite its size, Lewes was also included. However, we were concerned that a number of the other areas we had proposed for inclusion still lay outside the draft boundary. Consequently, we needed to engage with local groups to make a further case for these.

We also needed to consider the proposals for the administrative arrangements. We were broadly content with what was proposed in relation to planning: the National Park Authority to have control of planning; three joint structure plans and three joint mineral and waste plans; and development control possibly to be delegated to local authorities.

Crucially, we had to raise public awareness to ensure as many responses were sent to the Countryside Agency as possible. Chris Todd once more put pen to paper and produced another newsletter in December 2001 to alert our member organisations to what they needed to do to respond to the Agency's consultation. The culmination of the Campaign's work was our formal response to the Countryside Agency. 150 pages long, it was certainly a thorough piece of work, and one I was proud to put my name to. In March 2002 I was able to report to our Executive that I was very pleased with the last three months of intensive campaigning.

The Countryside Agency Board considered the results of the consultation at their meeting on 18 April 2002. The report to them recorded that over 7,100 people had visited their road shows. 6,676 had responded. Less than 4% of the responses opposed the National Park in principle.

Seven local authorities objected to the National Park in principle and only two supported it. The others did not state a clear view. With the exception of Winchester City Council, who wanted the existing two AONBs managed separately, and Mid Sussex District Council, who wanted the current arrangements to continue but with increased funding, those who opposed wished to see a South Downs conservation board.

For the first time there had been a consultation on a draft boundary. Importantly, given what was to come the Countryside Agency wrote:

'The 1949 Act refers to the importance of having regard to the character of the landscape. Landscape character assessments have revealed the complexity and variety of landscape character that exists within the South Downs. According to the Agency's approach to identifying national park boundaries, Areas to be included may be of differing landscape character: <u>quality will be the</u>

key determinate rather than uniformity (emphasis added)

A variety of landscape character is an important factor in the overall unity of the park. Usually however there will be some unifying factors, such as land use, ecosystems, historical and cultural links which bring different character areas together to be included in the park.'

At their next meeting the Board members expressed concern over the exclusion of three areas in particular: Woolmer Forest, Ministry of Defence land lying to the north of Liss in the so-called 'A3 corridor'; Kirdford and Plaistow, to the east of Petworth; and Ditchling, a downland village to the north of Brighton. There was hope, at least, for them.

In September 2002 Chris Todd drew up a paper which summarised achievements in the past seventeen months: we had expanded the size of our network, with more and more groups joining the Campaign every month. Our membership now comprised 52 organisations. We had distributed nearly 100,000 newsletters informing people about the national park process and the public consultation. We had released nearly 30 press releases and had written many letters to local papers raising awareness of the vulnerability of the South Downs environment and the benefits of a national park. Our representatives had spoken at numerous parish council/village society meetings and events. Chris Todd had acted in an advisory/consultancy role helping people (via email, telephone, letter and in person) understand the issues and what they need to do to make their case, either to their local authority or the Countryside Agency. We had produced a comprehensive case for additional areas to be included in the South Downs National Park, in a submission of over 100 pages, with supporting information running into many hundreds more.

There was a lot more to do.

CHAPTER ELEVEN

THE STATUTORY LOCAL AUTHORITY CONSULTATION AND THE DESIGNATION ORDER

Now that the Countryside Agency's public consultation was over, they were poised to begin their formal consultation with the local authorities as proposed in their latest timetable.

April 2002	Countryside Agency Board to consider results of public consultation and the boundary and administrative arrangements
Late May – June 2002	Statutory Local Authority Consultation on revised proposals
Autumn 2002	Draft Designation Order sent to Secretary of State, thereafter 28 days for formal objections
Late 2002	Decision on whether to hold Public Inquiry (automatically triggered if any Local Authority objects)
Spring 2003	Commencement of Public Inquiry
2004/2006	National Park Authority set up if Secretary of State confirms the Designation Order

Slightly later than planned, the statutory Local Authority Consultation was launched in Petersfield on 11 June 2002 and was scheduled to last until August 16. The Agency had responded to its public consultation by making some minor changes to the draft boundary and administrative options which it presented that day, together with its plans for the next stages in the designation and consultation process. The event was a useful one for us as we acquired several new member groups.

Given the numerous occasions on which the Countryside Commission, and latterly the Countryside Agency, had questioned the need for a South Downs

National Park and promoted a Conservation Board instead, we read their consultation document with great satisfaction:

'It would be possible to set up a Conservation Board under new legislation provided for by the CROW Act 2000. Some responses (to the public consultation) suggested it would be an appropriate way forward. However, while this Board would have permanent legal standing, it would not have the enhanced powers of a national park authority and would only have an advisory role in planning. It would also attract less funding than a national park authority because AONBs are funded on a different basis and at different levels to national parks.'

It was perhaps unfortunate that Defra chose to make a Ministerial announcement about its 'Review of English National Park Authorities' with a great fanfare at the CNP conference in Southampton on 13 July, whilst the South Downs consultation was still underway. This Review was wide-ranging with 54 recommendations, not all of which subsequently saw the light of day. As some of the recommendations required primary legislation, should they be taken forward, a consultation was launched to last until October. Amongst the recommendations were several related to governance and funding which interested greatly those politicians, both national and local, who were opposed to the National Park, because they smacked of increased central government control. The first of these was that the nationally appointed members be increased from a quarter to two fifths whilst the maximum number of members would be capped at 25. Consideration should be given to an independent chair and, finally, that the whole of the national park grant should be provided directly by central government rather than a quarter of it being channelled through local authorities.

There also appeared to be confusion about the nature of the South Downs consultation itself. It was not a public one on the principle of designation but one confined to local authorities essentially about the boundary and the administrative arrangements. Andrew Tyrie, MP for Chichester, who continued to be firmly opposed to the National Park, published a briefing note which suggested that the consultation process was flawed because it did not address the principle of designation and that there should therefore be a local referendum. In response the Countryside Agency had to issue a correction:

'The nature of the consultation reflects the fact that the Countryside Agency has decided in principle that there should be a national park and that issues on which views are now required are the practicalities. Thus the consultation did not ask a direct question on whether there should be a national park. The

Countryside Agency has a statutory duty to designate national parks and must use the statutory criteria which are laid out in the National Parks and Access to the Countryside Act 1949. The desire for a national park is not one of the statutory criteria and so there is no basis for the Agency to consult on whether or not there should be a park. However, the Agency has chosen to involve local people and organisations in the designation process.

The consultation was not therefore a referendum into whether or not there should be a national park, but an opportunity for people to give views and advice on both boundary and administrative issues. If people remain opposed to a national park at the end of the designation process, they can object to the designation order. They will also then have a chance to air their views at any Public Inquiry.'

We had no formal part to play during this consultation period but there was much to do. The consultation not only included county, unitary, district and borough councils but town and parish councils as well. By now we had a growing number of the latter included in our membership and many of them wanted additional land in their parishes included in the boundary. Members of our Executive, as well as Chris Todd, found themselves providing advice and support accordingly, not only to the parish councils but also to local amenity groups who wanted to lobby their councils. But the main thrust of our activity was the three areas that the Countryside Agency Board had indicated they wanted to consider further: Kirdford and Plaistow, Ditchling and Woolmer Forest.

Kirdford and Plaistow lie to the east of Petworth in the Western Weald. The Countryside Agency's consultants regarded this remarkably unspoilt and tranquil area as lying within what they described as a 'transition zone', that is where the quality of the landscape changes gradually and the precise point at which the boundary is drawn is a matter of judgement. The Countryside Agency had omitted it because they felt it lay beyond the point at which the quality of the countryside merited national park status and because they felt it had greater affinity with adjacent Surrey than with the South Downs. I, and other members of our Executive, worked closely with the Kirdford Conservation Society and the Plaistow Village Trust as they put together a robust case, helped by both societies having planning QCs amongst their members. Rather predictably in our view, Chichester District Council made no comment on the area and West Sussex County Council opposed its inclusion.

Ditchling is a quintessential downland village to the north of Brighton, nestling under the north-facing scarp slope of the Downs. Many artists and writers were attracted to Sussex downland villages in the first half of the Twentieth Century and Ditchling was particularly notable for its large community centred on the artist-craftsman Eric Gill. The Parish Council were very keen to see it included and we agreed wholeheartedly with their case. Chris Todd worked closely with them and others in the village to make the case for inclusion.

Woolmer Forest is owned by the Ministry of Defence and used as a training area. The southern part, lying to the south of the A3, known as Longmoor, had been included in the draft boundary. The larger area to the north of the A3, and ecologically the most important, had been excluded on the basis of somewhat questionable information provided by the Ministry of Defence with regard to public access. Margaret Paren, in particular, worked with parish councils and local groups to put together the case for inclusion. We were not alone in our quest to get this area included: Margaret contacted the late Sir Martin Doughty, then chair of English Nature, and was reassured that their submission to the Countryside Agency would support inclusion.

As well as all of this activity, we encouraged our member groups to lobby their local authorities and ran an accelerated media campaign to rebut what we regarded as the inaccurate statements from West Sussex County Council. In this we had much support from the local press in West Sussex which firmly supported the National Park.

After the consultation was complete, we were delighted to host Kate Ashbrook, still a member of the Countryside Agency Board, for a day-long visit to these areas, as well as others we were championing, so that she was fully informed for the forthcoming Board meeting. She agreed with our arguments, and we understood she had written to other Board Members accordingly. Later, the whole Board visited the priority areas for inclusion.

Meanwhile, our national organisations continued to play their part in ensuring the South Downs remained firmly on the national political agenda. Emily Richmond, the Ramblers Association's representative on our Executive recalled:

'One of the strongest campaign tools that the Ramblers' Association employed was taking MPs on walks on the last day of the political party conferences. This was relatively straightforward to organise when they occurred in Brighton, and on several occasions we gathered at the pier to board a coach, leaving behind the political wrangling, and enjoy some great views and fresh air. We used the walks to talk about important issues of access land, footpaths, encroaching development, rich and diverse flora, and the need for landscape protection. The

walks attracted a wonderful mix of MPs who, to varying degrees, cared about the campaign, loved a bracing walk and a free pub lunch, or just wanted to escape the cameras and hacks back at the convention centre. Clare Short, Chris Mullin, and Lord Rooker (minister for Housing and Planning) were some of the big names who came along. I was always impressed by their tolerance of our lobbying, taking every chance we could to elbow in another issue that they should be considering, and making the most of our 'in situ' platform to illustrate our points. Those walks with the MPs often resulted in rich conversations where, as campaigners, we really felt we could get to the heart of issue.'

The decision to proceed with designation or not now lay with the Countryside Agency Board and they were due to make their decision at their headquarters in Cheltenham on 14 November 2002 at a meeting starting at 9.30am. Some of our Executive resolved to go to Cheltenham to watch the debate and, hopefully, to be present at the historic moment when the decision was taken to proceed.

Margaret Paren arranged to travel with Owen Plunkett from the Hampshire Ramblers. The night before had been a particularly stormy one and, as she and Owen set off in the dark, there was still heavy rain and a high wind across Hampshire. So the first part of their journey was spent driving carefully along sometimes flooded roads, avoiding tree debris as they went. Consequently, they arrived at the car park in Cheltenham at just about the time the Board meeting started and ran to the Countryside Agency offices. To their chagrin, as they entered the boardroom, the first people they saw were some councillors and officers from East Hampshire District Council who had clearly made a better fist of the journey!

Meanwhile John Templeton, at the time still living in Ealing, left home at 5.50 in the morning, encountering wet and windy conditions on his way. He recalled the weather was so bad that it was announced on the radio that the Severn Bridge was closed, although this hardly affected his journey to Cheltenham.

The turbulent journeys were worthwhile. The Agency's Board unanimously agreed to proceed with designation and afterwards they were kind enough to share their coffee and biscuits during a break in proceedings. The journey home was undertaken in much better weather. John Templeton celebrated with a cream tea at Burford. Margaret Paren spotted a rainbow over Butser Hill as she drove along the A3. At the time she thought it presaged well for the rest of the process little realising what was to come.

On 18 December 2002 the Designation Order was made and in January 2003 it was put on deposit for 28 days thus allowing objections to be lodged to the principle of designation as well as to parts of the boundary. Within this time span it was essential that we persuaded as many people as possible to write in support of the principle and to lodge objections to boundary exclusions.

With the Designation Order the Countryside Agency published the revised boundary and the reasoning for it. All in all, there were 56 revisions. We were generally pleased with the results and delighted that Woolmer Forest and Ditchling had been included. However, we were disappointed that Kirdford and Plaistow were not.

Defra received 5735 responses, just over half of which were standard letters from Brighton & Hove Albion Football Club supporters anxious to ensure their new stadium was built, and other specific area campaigns. These letters were not individually written and, in the case of the football supporters, often arrived together in big boxes, so they were not even individually sent. These were discounted as not duly made representations. Of the remainder, which were all individually written, 72% supported the principle of a national park, while only 5% opposed it. Of the fifteen local authorities directly affected, only six still opposed it in principle: West Sussex County Council, East Sussex County Council, Chichester District Council, Mid-Sussex District Council, Wealden District Council and Winchester City Council. Three formally supported the principle: Brighton & Hove City Council, Lewes District Council and Hampshire County Council. Eastbourne Borough Council made a public statement that it supported in principle. East Hampshire District Council formally opposed the principle, but released a public statement explaining that this was a tactical move to gain a seat at the table at any forthcoming Public Inquiry when administrative matters would be discussed, but that it supported the National Park.

The Sussex Downs Conservation Board made no comments on the boundary.

At this stage, our Executive faced a conundrum regarding the boundary. We had to prepare for the likelihood of a Public Inquiry which would consider proposed changes to the Designation Order boundary. We were under pressure to support many additions to the boundary from numerous member groups. We recognised that although many cases had merit, not all were likely to meet the criteria for inclusion as set out by the Countryside Agency. We could therefore lose credibility if we supported all but could disappoint members if we did not do so. We therefore decided we needed a robust and transparent process for choosing those cases we would support.

Consequently, we decided to resurrect our boundary group with the remit of considering each and every case for inclusion and undertake an on-the-ground survey of the areas concerned. They were to report back their findings to the Executive and, once the Executive were satisfied which areas were deserving of support, this should be presented to all the member groups.

© *Chris Todd*
The Campaign's boundary group take a break during their travels
(From left Margaret Paren, John Templeton, Owen Plunkett, Paul Millmore)

This time the boundary group comprised: Paul Millmore, Margaret Paren, Owen Plunkett, John Templeton and Chris Todd. Other members of the executive with special knowledge of particular areas joined as and when necessary. Margaret Paren became the de facto chair of the group on the basis that as by far and away the majority of candidate areas were in Sussex, and she lived in Hampshire, she could be more of a neutral chair than the others.

John Templeton remembers a day-long meeting at Margaret Paren's house in early December 2002 at which they charted their plan of action. They decided to begin work in early January 2003 and foresaw the need for at least four days of site visits to complete the task. Unfortunately for them, the winter of 2002/2003 was Siberian, but they fortified themselves with flasks of coffee provided by Margaret, and by pub lunches of varying quality. Here, Paul conducted his own survey of the local beer and steak and ale pie.

The first area they considered was a small triangular field to the south-west of Storrington. Having peered over the hedge they decided a better viewpoint would be from Kithurst Hill on the South Downs Way, where they could view it from above. Margaret Paren recalls there followed several hours of quite heated discussion, with Chris Todd manfully dragging the conversation back to the Countryside Agency criteria whenever it threatened to become too philosophical. Eventually they decided to recommend to our Executive that the inclusion of the area should be supported. For the record, that field is now part of the South Downs National Park.

The group realised that, at this rate, the job could take much longer than planned but as time went by, they found that consensus was reached ever more quickly, and by the end they barely needed more than a few minutes to agree their recommendation.

The group was joined from time to time by proponents of the area in question. Much the most memorable of these encounters was when Dave Bangs joined them between Newhaven and east Brighton. His knowledge was breathtaking but so too was his disregard for the Countryside Agency's criteria. John Templeton remembers a bitterly cold morning on a hill above Whitehawk in east Brighton when Dave lambasted Chris Todd and Margaret Paren about their adherence to the application of the criteria with Dave loudly proclaiming that a national park should not just be one for cream teas but fish and chips as well.

The Campaign's Executive robustly debated the group's findings and on more than one occasion they were sent back to revisit the area in question. On one occasion, regarding the area to the south of Arundel, they were sent back, very reluctantly, a third time.

Finally, the job was complete, and Margaret Paren presented the recommendations to representatives of our member groups at a half-day conference held at University of Sussex in October 2003. There was obviously disappointment from groups supporting areas we proposed not to endorse. But we made clear to them that our decisions were non-binding on them, and they were free to pursue their own areas if they wished to do so. We would not contradict them or argue against them in public. On this basis the meeting agreed the findings. We thus had a sound basis to proceed and a united Campaign.

CHAPTER TWELVE

THE INQUIRY: FIRST PHASE

It was apparent that a Public Inquiry would be necessary given it required only one local authority to object to the principle of establishing a national park to trigger one. At a meeting in May 2003 the Countryside Agency confirmed this was the case.

With regards to the boundary, eight local authorities, over twenty parish councils and thirty individual landowners and agents called for deletions of areas. Fifteen developers, Southern Water and Portsmouth Water were also calling for areas to be deleted from the designated South Downs National Park. Altogether forty areas were being contested, not including those proposing a 'chalk-only' national park that would exclude the whole of the Western Weald.

On the other hand, there were nearly ninety cases calling for additions to the boundary coming from local authorities, parish councils and campaigning groups. Altogether it was estimated that there were up to 180 full cases and that, consequently, the Inquiry was expected to last twelve to fourteen months. We were advised that the Inquiry would open in Worthing in November and that a pre-Inquiry meeting would take place there on 7 July.

However, much the most disturbing news for us was that the Inquiry would be conducted along the same lines as a Local Plan Inquiry and, although the exact format was in the hands of the Inspector to decide, it was likely that only objectors would be able to give evidence. This meant that we would not be able to support the principle of a national park.

The pre-Inquiry meeting took place as planned on 7 July. Mr Robert Neil Parry introduced himself as the Principal Planning Inspector appointed to hold the forthcoming Inquiry and to report to the Secretary of State with recommendations. We later learnt that he was a very experienced Inspector chosen, in part at least, because he was a 'people person' thought well able to cope with the numerous representations from frequently differing perspectives. And so it proved: a lean, grey-haired Welshman with a neat goatee beard, he almost always remained affable and good-humoured as he navigated his way through often conflicting evidence. The result was that though the arguments were often robustly made they seldom became acrimonious.

The terms of reference for the Inquiry were set out. There were four issues to be addressed. First, whether the area, as a whole, met the criteria and purposes of designation of a national park as set out in the 1949 Act. Secondly, whether the proposed boundary was properly drawn, bearing in mind the criteria and purposes of a national park. Thirdly, was a new national park authority the best means of managing and administering any new national park? Finally, there was the question: should the East Hampshire and Sussex Downs AONBs be revoked?

For us, the main news was that though oral evidence would be confined to objectors, the Inspector was prepared to receive written representations in support of the principle of designation. Armed with this, we decided that we must submit an opening statement that set out our pitch and which, hopefully, might influence the debate that would follow. Paul Millmore was particularly keen that we highlighted to the Inspector what he was unlikely to hear from opponents.

Work began on preparing it. As part of this evidence, our first major salvo in a long line of documents, we detailed the strong public support for the National Park, why the South Downs met the criteria and how previous designation decisions had been flawed. We moved on to describe the South Downs as the most iconic of lowland landscapes, symbolising England for residents and visitors alike. Few other regions, we suggested, had the same emotional pull, that quality of remoteness, wildlife and cultural heritage that distinguished the chalk Downs and Western Weald from other lowland landscapes.

We continued by detailing the qualities of the rippling chalk hills, the magnificent coastline culminating in the Seven Sisters, the mysterious Western Weald, the Hangers made famous by Gilbert White, the river valleys and rare chalk rivers and streams and, finally, the coastal plain. We explained how they combined to provide exceptional recreational opportunity. And we emphasised the proximity of the South Downs to urban centres: the strong network of rail and public transport links in the crowded south-east of England.

This document also addressed why national park status was especially desirable and how the current conservation model, while having had some modest successes, was not up to the job of safeguarding the South Downs for future generations. To illustrate the point, a number of newspaper articles were included from The Argus (The Evening Argus at the time) questioning the independence of the Sussex Downs Conservation Board. These were supplemented with numerous opinion pieces in the West Sussex press backing the National Park and calling on West Sussex County Council to drop its opposition.

One paper we commissioned in support was 'The Specialness of the South Downs' by Dr Peter Brandon, who had served on our Executive in the early days of the Campaign and was the former head of the Department of Geography at the University of North London. He was a native of the South Downs and had spent many years researching the history, culture and geography of the area which he presented in his splendid book 'The South Downs', published in 1998.

Margaret Paren tasked me to write another entitled 'The Geology and Biodiversity of the Designated South Downs National Park - An area of national and international importance'. I commissioned Dr Rendel Williams, reader in Physical Geography at the University of Sussex, to provide the geological section and set to work myself to write the section on biodiversity, relying heavily on the Sussex Wildlife Trust and the works of Dr Francis Rose.

The work of the boundary group ensured we had a clear idea of what we wanted to put to the Inquiry. What we needed now were the Proofs of Evidence to submit to it. Chris Todd produced a template to be used in all cases and a whole phalanx of authors set to work. He, Margaret Paren and Ruth Chambers read through all the drafts: first, to ensure consistency so that one Proof did not contradict another, and secondly to question any unsubstantiated statements that could be challenged at the Inquiry. At this time desktop publishing was in its infancy and contact was via phone lines that were less than reliable. The work of assembling the proofs fell to Chris Todd and this involved checking and attaching much supporting evidence, including maps and photographs, and often had to be done manually. Chris was assiduous in ensuring that everything was in place and properly indexed. We therefore felt well prepared for the Inquiry itself.

As the proceedings continued, month after month, so did the Campaign's number of submissions increase. It is only possible to focus on the Campaign's principal contributions within the confines of this book. But suffice to say that by the end we had presented over eighty Core Documents and Proofs of Evidence for this part of the Inquiry.

Before proceeding further with our South Downs story, I should add a word or two about the designation process for the New Forest National Park. This had proceeded ahead of that for the South Downs and by the time our Public Inquiry started theirs had finished and the Inspector's report was awaited. Although their Public Inquiry had been conducted by a different Inspector, a number of the main players attended both inquiries, often representing different clients.

On 10 November 2003 the South Downs Inquiry duly began at the Assembly Hall in Worthing with the Countryside Agency setting out the main bones of the case for a national park. On the second day we all moved to the more permanent venue, the cosier environment of the Chatsworth Hotel, also in Worthing. We were to spend many, many months in a ground floor room in this hotel, just a block from the sea and overlooking Steyne Gardens, a rectangle of welcome green space in an otherwise urban area. The room itself seemed something of a throwback to earlier, more genteel times: its rather drab décor enlivened by gold-framed and red velvet chairs for those of us sitting in the audience.

Here things livened up when the main opponents of a national park, West Sussex County Council and Chichester District Council (the councils), with a budget of £200,000 for fighting their case, began their attack on the Countryside Agency's proposals. Though this was not an unusual amount of money for mounting such a challenge, it seemed an enormous sum in contrast to our own much more limited resources.

Their main arguments were all too familiar by now: that a national park had already been rejected three times; the area had a limited sense of wildness and remoteness; and it had far less public access than any other national park. They also claimed that the Countryside Agency had misapplied and misinterpreted their own designation criteria and that their consultation process had not been conducted properly. They argued that a conservation board under the CROW Act would be better and more democratically accountable and would not be bound by the membership requirements for a national park authority. It would be preferable because it would not have to promote recreation, which would be beneficial for the protection and management of a fragile landscape. Planning would be more effectively carried out by the existing planning authorities. Finally, they argued that if there was to be a national park, it should be confined to the chalk ridge of the South Downs, a view which they pressed relentlessly until the end of the Inquiry.

In total the councils spent over two weeks fighting against the principle. Their legal team was led by Rhodri Price Lewis QC, another Welshman, assisted by Scott Lynees, and two officers from West Sussex County Council gave evidence. Their expert on landscape was Moira Hankinson who had appeared at the New Forest Public Inquiry against the Countryside Agency and who, again, unceasingly challenged their findings.

Throughout the full length of the Inquiry the Campaign had representatives present almost every day. By contrast we never identified a single West Sussex

County or Chichester District Councillor in attendance despite their councils being such vociferous opponents and spending much public money on their opposition.

Set against them, the Countryside Agency was represented during this first phase of the Inquiry by two of their officers. Their legal team was headed by the ebullient Robert Griffiths QC, completing the triumvirate of Welshmen.

The Countryside Agency's choice of landscape experts caused us ever-increasing concern. The Agency had employed Landscape Design Associates as their consultants up to this point. That team of consultants had been headed up by Professor Robert Tregay. However, for the Inquiry, the Agency had run a competitive tender and, as a result, had switched consultants. This meant that the new team, headed by Martin Leahy, had to defend decisions they had not made themselves and worse, Martin himself had appeared at the New Forest Inquiry for objectors and vociferously opposed the Countryside Agency. Thus, when cross-examined by Rhodri Price Lewis, he was forced to dance on a pinhead to defend what he was saying now against what he had said then. He showed considerable skill at this wordplay, but we felt that, consequently, the case for the South Downs got somewhat sidelined.

Winchester City Council, Mid-Sussex District Council and Wealden District Council then joined forces to give somewhat less strident evidence than the two main opponents and their case only lasted for a morning. East Sussex County Council was next up, and its evidence was given by Steve Ankers, then an officer of the Council, later employed by the South Downs Society (the renamed Society of Sussex Downsmen) before his untimely death. He focussed on the process rather than opposition to the principle. There then followed the National Farmers Union, the Country Landowners Association, and a number of individuals. The Conservative MPs Howard Flight (Arundel and the South Downs), Andrew Tyrie (Chichester) and Nigel Waterson (Eastbourne) also objected. There were objections from seven Parish Councils.

Once the 'in principle' hearings were over, in January 2004 the Inquiry moved on to consideration of the boundary.

The discussion on the boundary took each sector in turn, hearing cases for inclusions and exclusions. It was clear that the Inspector enjoyed hearing the evidence from local people, though he initially looked shocked when Professor Robin Milner-Gulland, one of our Executive members, opened his attaché case and dramatically pulled out a substantial piece of greenery in support of the case for a relatively small triangle of land between the re-aligned A283 and Washington village. The ruse worked: the land was included.

During this stage of the Inquiry, the Campaign faced a dilemma in putting our case: if we challenged the Countryside Agency's landscape consultants too rigorously, we undermined their standing with the Inspector, if we did not press hard enough, we risked losing the additional areas we wanted to see included. However, we decided that, given the huge clamour from member groups and the public for more areas to be included, we had to take the risk and challenge the boundary.

We focussed on providing information on the boundary criteria, especially where we had new evidence, as this would give the Countryside Agency a way of accepting our proposals without losing face. We tried to avoid challenging them head on, but our frustration was the Agency's defence of the sometimes indefensible. In these situations, we had no choice but to call into question their evidence. One of the worst instances arose when we proposed the inclusion of more land in an area of the Low Weald to the east of the Adur Valley, including the picturesque Sussex Wildlife Trust headquarters and education centre at Woods Mill. Here, in arguing their case, the Agency submitted pictures of landscape detractors that they claimed showed why the area should not be included. We were able to show that these pictures were outside of the area in question. We won our case but in doing so had had to lay open the poor quality of some of their evidence.

Mention should also be made at this point that the Inspector, like his counterpart in the New Forest Inquiry, had his own expert landscape assessor. At the time he was employed by Scottish Heritage, which was the Scottish equivalent of the Countryside Agency. Again, this followed the precedent set by the New Forest Inquiry. He attended many Inquiry sessions alongside the Inspector. Through later Freedom of Information requests, we determined that he was given unpaid leave to attend the Inquiry, though this was not adequate for all the work he undertook, which suggested some had to be completed in his own time. He introduced much controversy later in the process and in describing this later I will refer to him as the Landscape Assessor.

This stately progression around the boundary was scheduled to be interrupted by evidence from the councils on their proposal for a 'chalk-only' national park. On the day that evidence was to be presented, the New Forest National Park Inspector's report was published. Although he recommended a national park, he cut large chunks out of the designated boundary, significantly reducing the size of the proposed national park, and leaving out the main settlement of Ringwood. The Inspector's rationale for such a major change was that land included must be of New Forest character and he had therefore removed land along the Avon Valley and elsewhere that did not meet this criterion.

This was a stunning and unexpected blow. Although the Scottish National Park legislation of 2002 talked of distinctive character and coherent identity for national parks there, no such test existed in the 1949 Act. Where had this come from? Perhaps, we speculated, because the New Forest Inspector's landscape assessor, like ours, was from Scotland? As we trooped into the Inquiry room, we gloomily reflected that this made the Western Weald and the towns of Lewes and Arundel decidedly vulnerable in the case of the South Downs.

At the start of proceedings that day counsel for the councils asked for a postponement so that the implications of the New Forest Inspector's report on the arguments being presented on the South Downs could be considered further. Mr Parry acceded to the request.

This at least provided us with the time to decide what we could do to help prevent the outcome we feared. We decided we needed to produce two further documents: one to refute the single character argument and the other in defence of the inclusion of market towns.

It is fair to say that the former was easier to write. There was a wealth of information to demonstrate that most of the national parks in England and Wales encompassed more than one type of geology and landscape type and that Dower and Hobhouse had deliberately proposed this be the case because contrasting and complementary landscapes provided a greater variety of recreational opportunity. We were indebted to Donna O'Brien, who had joined CNP, and who took the lead in writing this important paper: I can do little more than quote here the concluding paragraphs.

'When Dower and Hobhouse were writing their reports, they made a deliberate attempt to include different landscapes within the geographical unit of a national park, since it was the scenic quality, not scenic uniformity, which defined their thinking. Without this approach the national park family would be much poorer, Northumberland would be confined to the Cheviots, Pembrokeshire to the coast, and Snowdonia to the Snowdon massif. The Peak District would have to have been split into more than one national park, while the Lake District would be without Lake Windermere. Gathering an assemblage of landscapes for inclusion in the South Downs National Park is not unusual, the approach was pioneered over fifty years ago and is as valid today as it was then.

Restricting the National Park to the chalk ridge not only severely reduces the extent of the Park but it also restricts the variety of recreational opportunity, and conflicts with existing government guidance which recognises that parks "contain a variety of landscapes, capable of accepting and absorbing many different types of leisure activity".

In conclusion, the South Downs Campaign submits that what marks out the national parks from the surrounding countryside is not their different character but their high and distinctive quality which is a reflection of a wide variety of factors, as the Dower and Hobhouse reports so clearly stated. In the designated South Downs National Park, the natural beauty and opportunity for enjoyment comes from the combination of scenery. The complementing upland and lowland, the more intimate western Wealden landscapes, the river valleys dissecting the chalk and the expansive views and openness of the Downs, particularly in the east, and much more, in short 'the exquisite arrangement and combination of the whole.'

The paper on market towns was more challenging. Petersfield, with a population of around 14,500 was the largest town in any AONB in England or Wales. Lewes was even larger, with a population of around 16,500. By contrast, the largest towns in existing national parks, Brecon and Windermere, had a population of under 8,000, not much larger than Midhurst.

The Countryside Agency had drawn up criteria for the inclusion of towns in a national park which amounted to: consideration of the quality and historic value of the built environment; the quality of the surrounding countryside; its gateway potential and its economic potential. They had made objective assessments in relation to each of the towns included in the Designation Order boundary and it was difficult to see what we could add, so our paper re-iterated that there were no statutory or policy constraints to the size of settlements in a national park and provided the Inspector with more detail about the quality of the towns in question. We were conscious that there would be an attack on the planning load created by their inclusion but countered this by suggesting inclusion or otherwise should be determined on a case-by-case basis: the decision should be based solely on the statutory criteria and planning was not one of these.

The discussion on the 'chalk-only' national park was rescheduled at the end of this phase of the Inquiry, along with that on the inclusion of Arundel, which West Sussex County Council was also pressing to be excluded.

The shape of things to come had been presaged in a resolution passed by the Sussex Downs Conservation Board in January 2001 in response to the publication of the Area of Search. It read:

'This Board views with concern the substantial area of land which is

essentially Wealden in character and which does not offer a sense of wilderness recommended in the Edwards Report that is proposed for inclusion in the proposed South Downs National Park and requests the Agency seriously to reconsider its inclusion and whether the area meets the statutory criteria for national park designation.'

Moira Hankinson gave evidence on the councils' behalf. Her submission was based on her experience as a witness in the New Forest National Park Inquiry and on the decisions of its Inspector. She had argued at that Inquiry that the critical test for boundary making was the presence of 'New Forest character'. That Inspector's landscape assessor had suggested that the majority of the Avon Valley, though outstanding in its own right, did not possess sufficient cultural, ecological or land management connectivity to justify the valley landscape being part of the suite of types that formed the distinctive landscape that is the New Forest. The New Forest inspector had concurred.

By applying the same test to the South Downs National Park, it would exclude much of the Wealden and Coastal Lowland areas: these did not constitute the landscape of the South Downs, nor did they display a firmly established South Downs character. Furthermore, she argued that the Western Weald had more in common with the Surrey Hills AONB than the South Downs and that the Countryside Agency had not proved that the areas provided a recreational experience superior to that of other AONBs; nor were there extensive areas of 'relative wildness'.

The Countryside Agency's main rebuttal on the 'chalk ridge only' alternative was that it was not a duly made objection since the councils had not presented the Inspector with a boundary. As such, the Agency also said that it was irresponsible of the councils to have brought the 'chalk-only' argument forward like this, which was really another in principle objection. We were pleased that the Agency made extensive use of our papers on the landscape of other national parks and market towns in refuting the councils' arguments.

This issue of a lack of a boundary is an important one in terms of what happened later in the process so is worthy of an explanation here.

In the New Forest Inquiry, a number of objectors had put forward alternatives to segments of the boundary, leaving just a few small gaps to be drawn in to produce a workable though heavily modified boundary. The Inspector had therefore felt able to present to the Secretary of State an alternative boundary that met his adopted criteria for boundary setting.

This was not the case in the South Downs. The councils did not provide an alternative boundary, they produced instead an 'Area of Search' within which, they suggested, a new boundary could be drawn should the Inspector accept their arguments for a 'chalk-only' national park. This approach was probably adopted because if they had produced a boundary, the Countryside Agency would have been able to challenge it step by step. It also overcame the difficulty that part of the area they wished to exclude was in Hampshire and supported for inclusion by both Hampshire County Council and East Hampshire District Council.

When the map of their Area of Search was published the then Leader of East Hampshire District Council, Elizabeth Cartwright, met with the then Leader of Hampshire County Council, Ken Thornber. Much to our delight he instructed that a counter objection should be drawn up. This was done but turned out to be a puzzlingly low-key document, a disappointment to both the Campaign and the Countryside Agency.

In December 2004, over a year after the opening of the Inquiry, we delivered our closing statement on the boundary. Altogether, we had presented evidence on twenty-four different areas including ones on the Kirdford and Ifold area, and on an open boundary to the sea. After summarising our technical arguments, we concluded with two significant points:

'Most previous national park designation processes have been contested in one way or another, often by local authorities fearing loss of their power. All are now accepted, indeed supported, by many of those who offered opposition at the outset.

To some extent the South Downs National Park designation process has followed this same pattern, though the extent of local authority opposition has been remarkably limited given the number of authorities affected in this case. But it is difficult to name another national park designation process during which there has been such a groundswell of popular support, both from those living within the designated boundary and from outside. And this popular support shows no signs of slackening, indeed it is growing. During the course of this Inquiry, the South Downs Campaign's membership has grown by over 20%. If confirmed, the South Downs will be the most visited national park in the country, proof that national park status for this most popular and iconic of English landscapes is well deserved after a fifty-year delay.'

After the Inquiry had completed its long and exhaustive examination of the boundary it turned to consider the most appropriate way of managing the designated South Downs National Park.

For us, the most crucial session was that on planning. Strategic planning, as

I have alluded to previously, did not create any problems. We, the Countryside Agency and the local authorities were at one in believing that joint plans were the answer. Planning policy was a different matter. We had always been firm and united in our belief that the National Park Authority should have complete control over this, the local plan, as it was key to consistent decision making across the whole area and would ensure that the National Park Authority was in the best position to deliver national park purposes. We were therefore opposed to any solution that could result in a dilution of this principle. Here we supported the Countryside Agency though there were differences between us and some, at least, of the local authorities.

Development management, or control as it was then called, was more difficult for us. With an estimated 110,000 people living within the designated boundary, the South Downs would have a greater population than all the other national parks put together. The development management load would therefore be massive; indeed, today, the South Downs National Park Authority is one of the largest rural planning authorities in England. The issue was how this should be administered. The standard model in England was for all such decisions being taken by the relevant national park authority. There was a different model in the Cairngorms, whereby planning lay with the relevant local authority, with the national park authority having call-in rights for the most difficult and contentious cases. An alternative model had been tabled in the designation process whereby the National Park Authority would delegate decision making back to the existing planning authorities, the districts and boroughs, based on the policies it had included in its local plan. The local authorities had favoured this approach should there be a national park authority with planning powers.

This delegation model had been the subject of some debate within our Executive for some time but up to this point we had managed to avoid outright disagreement on it. But when it came to what stance to take at the Inquiry a decision had to be made. Both CNP and the Ramblers Association were reluctant to endorse delegation and preferred the standard model whereby the National Park Authority took all such decisions. They were supported by some others who were concerned that, if there was delegation, largely urban authorities like Brighton & Hove would continue to be in control of decision making related to their rural areas and their track record in this was not thought to be very good.

On the other hand, there was the significant risk that, if the National Park Authority took all planning decisions, the potential volume of planning applications might swamp the other work of the authority and, in the context of the Inquiry, give greater credibility to the West Sussex County Council argument for a 'chalk-only' option that excluded all the towns.

In the end after much debate, we agreed to propose to the Inspector that the National Park Authority should have full planning powers with a scheme for delegation back to existing planning authorities for all but the most important decisions. We recognised that the small volume of planning decisions within the national park in some local authority areas, like Brighton & Hove and Eastbourne, would probably mean that it would not make sense to delegate in those cases. We proposed that this arrangement should be reviewed after three years. However, for understandable reasons, neither CNP nor the Ramblers Association felt they could sign up to this solution. Thus, at the round-table discussion on planning, the Campaign found itself at one with the local authorities and opposed to two of its own key members.

Despite all the discussion about it during the designation process, there were still some doubts about whether delegation was or was not legally possible. It was an issue which had been around since 1999 when we heard that Defra had concluded it was not. This had led at that time to a tentative suggestion that it might be timely to look at a national park without the Western Weald, as this might overcome some of the alleged difficulties with planning. However CNP's pro bono barrister held firmly to the view that it was possible and, thankfully, never again thereafter was there any suggestion that the Western Weald would be sacrificed on the altar of planning.

The issue was finally put to bed by Robert Griffiths who advised the Inquiry that, although delegation as such was not possible, an agency arrangement could be made under the terms of the Local Government Act of 1972 which would amount to the same thing.

Alongside planning, we believed that much the most important issue to be addressed, in terms of administration, was how agri-environmental schemes for farming should be delivered in the future. At the time the Inquiry was taking place the most far-reaching reforms in the Common Agricultural Policy for fifty years were in train. This seemed to us to present a fantastic opportunity for a newly created National Park Authority to address the fragmented, uncoordinated and financially inefficient farming support then in place. We were fortunate not only to have the knowledge of Pat Leonard available to us but that of Ian Hill, who had been a principal agriculturist with the World Bank, and John Lomas, Director of Conservation and Development at the Peak District National Park. I joined this distinguished group and was actively involved in the writing of the resultant Proof of Evidence. We envisaged the National Park Authority providing the focus and mechanisms to conserve and enhance the National Park by addressing poor take up of environmental schemes, ensuring new schemes delivered jointly a better outcome and devising new and innovative ways of restoring and recreating landscape. We recognised the Authority would need the necessary power and resources to manage the new schemes in order to achieve these goals. It would be necessary for the scheme for higher level environmental delivery to be applied to the whole area of the South Downs. Defra staff responsible for the delivery and monitoring of the new schemes would need to be seconded to the National Park Authority, which itself would need to be encouraged to devise innovative solutions.

This was, as we recognised, a bold and far-reaching agenda, very much dependent on central Government support and agreement. Sadly, it was not to be. Whitehall clung onto its power and the Authority didn't receive the necessary powers or money.

Regarding the remaining administrative matters, we advocated that the Secretary of State should appoint more than the recommended twenty-two members. We did so because of the large number of local authorities involved. This was in line with the arrangements in the Peak District. We also made various recommendations regarding the streamlining of the delivery of countryside services and conservation programmes, with the National Park Authority acting as a facilitator and broker in bringing together the various agencies and users.

16 December 2004 was the last day of the public hearings. They had spanned 13 months. A total of 147 individuals and organisations appeared to make their case, many on more than one occasion. Almost all who appeared put in a substantial amount of work, often undertaken in their own time and at their own expense, and they submitted detailed evidence. Over 90% of the hearings were held in an informal manner, without the use of significant legal representation or expert witnesses, which allowed objectors and the Countryside Agency to discuss the issues with the Inspector in a more relaxed but no less demanding way. All of this was a reflection of the immense value placed on the South Downs by many people and organisations, whatever their views on the future management of the area.

The Inspector left open the Inquiry until after he had been on all his site visits, had read the written representations and clarified any outstanding points with the Agency. He therefore set the date of 18 March 2005 as being the closing day of the Inquiry. The final salvo was fired on behalf of the Countryside Agency by Robert Griffiths who accused West Sussex County Council and Chichester District Council of allowing their fear of the unknown and concern over their loss of planning powers to cloud their judgement about the case for national park status for the South Downs. He concluded that *'No lesser designation would be worthy of its true merits'*. In repost Rhodri Price Lewis claimed, on behalf of the councils, that no effective or convincing case had been made for national park status and that *'what was required was not a body with a statutory responsibility to promote further recreational opportunities, but one legally directed towards conservation of the landscape, a conservation board.'*

Through many years we were fortunate to have the local press on our side from The Argus in Brighton to the Petersfield Post and Petersfield Herald in Hampshire. But the most important, given the hostility of local authorities in West Sussex, particularly in Chichester District, was the family of newspapers covering that area, in particular the West Sussex Gazette. This paper carried a banner across its front page each week supporting the National Park. As the Inquiry drew to a close, I wrote a letter to the editor giving our heartfelt thanks.

The Agency issued a press statement saying:

'We do not believe that we have heard anything at this Inquiry to change our mind that all of the land we have designated fully meets the high standard required to be England's newest national park and that national park status is the best way of providing the protection and integrated permanent management that the South Downs urgently needs.

The Agency would like to thank all those who have contributed to the Inquiry. We believe the process has been a fair and open one. It will allow the Government to reach an informed decision as to whether the National Park should be confirmed, where its boundaries should be and how it should operate. It has also been a very positive process.

The sheer number of people and organisations who made the effort to formally support a national park clearly shows that there is overwhelming support for the confirmation of the South Downs National Park by the Government. We recognise that a minority of people and organisations still have concerns, but this Inquiry has allowed them to make their views known. We also believe that there was a consensus during the Inquiry that a future national park authority can operate effectively in the South Downs.'

CHAPTER THIRTEEN

MAJOR SETBACKS

In 2005 the whole basis upon which the Inquiry had spent many months was gravely threatened by an unexpected court case unrelated to the South Downs. Following the submission to the Secretary of State of the Inspector's report in relation to the New Forest National Park, and the subsequent signing of that Confirmation Order, there was a period within which a judicial review could be mounted against the Order. Taking advantage of this process, the Meyrick Estate submitted an application to the High Court to quash the New Forest National Park (Designation Order) in so far as it applied to the main part of the Hinton Estate, a large property totalling 6,000 acres (2,428 hectares). Mr. Justice Sullivan, a High Court judge, upheld their case. In doing so he interpreted the criterion for designation of 'natural beauty' extremely narrowly. He also implied that only areas with immediate recreational access, rather than 'potential access', could meet the recreational criterion. Various pronouncements were also made including that land in a national park should display a 'high degree of relative naturalness' although what this might mean in practice was not explained. The general implication and cloud of uncertainty cast by this judgment, which became known as the 'Meyrick Case' was that land clearly influenced by man should not be in a national park.

Since most existing national parks contain similar estates to the Hinton Estate (for example Chatsworth in the Peak District), as well as settlements, infrastructure, archaeological sites and managed farmland, to name just a few man-made elements, the ramifications of this judgment were potentially devastating. Perhaps the most striking thing about it, aside from the fact that the evidence never gave the judge the opportunity to consider the situation in the existing national parks, was the fact that no evidence was presented that considered the original reports into the establishment of the national parks, such as Dower in 1945 and Hobhouse in 1947, nor subsequent reviews such as Sandford in 1974 and Edwards in 1991, all of which made explicit how the national parks system in England and Wales was a product of man's influence on the landscape. Indeed, the National Parks Committee criteria for selecting areas of land for inclusion within the boundaries of national parks recommended including land of cultural, architectural and archaeological interest.

This judgment not only threatened the whole basis on which the designated South Downs National Park boundary might be cast, but also had serious implications for all the existing national parks in England and Wales. We were aware that Defra was planning to appeal the decision but were concerned that, if left to the courts, the outcome was not certain. We were therefore not alone in lobbying for legislative changes.

Defra consequently embarked on a twin track approach of applying to the Appeal Court and adding amendments to the Natural Environment and Rural Communities (NERC) Bill which, fortuitously, was passing through Parliament at that time. Two late clauses were introduced in the House of Lords which effectively took the designation criteria to where they were thought to be prior to Judge Sullivan's findings. The official opposition tabled an amendment to the Government's amendment seeking to embed the Meyrick Case judgment in law, so it carried significant weight. However, the late Earl of Selborne, the then influential Chair of the Environment Select Committee and brother-in-law of Minette Palmer, made a powerful speech in which he said that he would not support his front bench and would abstain. This ensured that the Government's amendment got through. Both amendments were passed at the Lords Report stage and the NERC Act was given Royal Assent on 30 March 2006. As well as containing the vital sections regarding national park designation criteria it also established Natural England, which subsumed the Countryside Agency.

The Meyrick Case was heard by the Court of Appeal on 1 November 2006. It was bitterly cold. I found myself sitting for the whole of the hearing with air-conditioning blowing icy cold air onto my neck. I was accompanied by a strong contingent of our members from the Campaign including John Templeton, Minette Palmer, Margaret Paren and Chris Todd. West Sussex County Council was also present.

The chair of the Court of Appeal said at the outset that they were somewhat frustrated just to be considering the case before them against the newly amended legislation rather than having the opportunity to make new case law. He conjectured whether to proceed or not. There followed a largely technical discussion on the merits or otherwise of the Meyrick Case itself. It was a great relief that the hearing was thus confined and did not stray into a more general discourse on the merits of the National Parks and Access to the Countryside Act 1949. The hearing completed, we realised it would be some time before we received the report on the Appeal Court's findings and thus know the future of the South Downs National Park designation process.

In these circumstances the Campaign found it challenging to plan its future. We still had no knowledge of the contents of the Inspector's report. We were also warned that some of the national organisations that were key members of the campaign were suffering financial difficulties that could impact on their support in the future. Somehow, we held together.

We focused our attention on local issues and awaited the Inspector's report with eager expectation. Meanwhile a considerable amount of discussion took place about the future. At the outset of the Campaign there had been an agreement that once a national park was secured the South Downs Society would become the de facto national park society. Representatives of the Campaign had several meetings with them to explore how they would carry out this role and what sort of engagement there would be with the many groups that were by now members of the Campaign.

Separately, moves were afoot to prepare for a national park in respect of the governance of the two AONBs. The Sussex Downs Conservation Board merged with the East Hampshire AONB's Joint Advisory Committee into a single South Downs Joint Committee. Representatives of all 15 local authorities having areas within the two AONBs were members, along with 13 appointed by Natural England. The body had a somewhat unwieldy membership of 45 under the chairmanship of Lord Renton. A number of our supporters were appointed to the new body by Natural England: Paul Millmore, Minette Palmer and Margaret Paren. Caroline Dibden, a stalwart of CPRE Hampshire whose professional background as a geologist became important to us later, was also appointed. As chairman of the South Downs Advisory Forum, Christopher Napier who also chaired CPRE Hampshire, was a co-opted member. The Advisory Forum was set up as a 'critical friend' to the Joint Committee and had over 100 organisations in its membership, including parish councils and landowners as well as conservation bodies, all represented by an executive panel. A number of members of the Advisory Forum's executive panel were, or became, members of our Campaign's Executive.

On the day the NERC Act received Royal Assent, 30 March 2006, the South Downs Public Inquiry Inspector's first report was delivered to Government. It took until February 2007 for the Court of Appeal judgment to be published. This upheld the Order of the High Court, although for technical reasons that differed from those in the original judgment. It in no way reinforced the findings of the lower court in terms of designation criteria for national parks. And so, on 2 July 2007, more than two years after the Inquiry had closed, the Inspector's report was finally made public.

We were deeply shocked, though in our hearts not surprised, that, although the Inspector gave unequivocal support for the National Park, and was complementary about the Campaign, he recommended that there should be a 'chalk-only' national park that excluded the Western Weald and all the main settlements, including Petersfield, Liss, Midhurst and Petworth. He also excluded Arundel and Lewes. In doing so, he noted that no-one had proposed an alternative chalk-only boundary and therefore extra work would be needed to identify a more focussed boundary. To help this process he provided a plan to illustrate the extent of the proposed new national park and sketched in an indicative boundary.

The Inspector did also recommend that the designated boundary should be extended by the inclusion of thirty additional areas, mainly on the chalk, and he also accepted our arguments for an open boundary to the sea, of which more later. However, he recommended twenty-four deletions including a substantial area of the 'Bentley Nib', which runs up the B3004 from Selborne to Bentley, covering the northwards extension of the Hampshire Hangers. Land round Coldwaltham, Steyning and the Adur valley to the north and the quintessential downland historic village of Ditchling were also excluded. With the exclusion of the Western Weald our case for the inclusion of the Kirdford and Plaistow area also fell.

Along with the Inspector's report, also published were two reports by his Landscape Assessor. The first related to evidence submitted to the Inquiry. But it was the second, a report on the so-called 'A3 corridor' in Hampshire and the farmland of the Lower Rother Valley in West Sussex, which concerned us more. It contained a welter of information that had not been heard at the Inquiry from any of the witnesses and contained major inaccuracies. Moreover, it claimed that the whole of the Rother valley, from north of Liss to Petersfield and across through Midhurst to Petworth, an area of over 100sq kms, protected as AONB since the 1960s, had been downgraded to such an extent by the cumulative impact of new roads, traffic noise, new and unsympathetic development and intensive farming practices, that it no longer met the natural beauty criteria which is the same for AONBs and national parks. This was especially alarming because, if accepted, the protected status of the area as an AONB would certainly be in jeopardy.

Short of dismissing the case for a national park the Inspector's report could hardly have been worse.

I recall a meeting on the Saturday morning following the publication of the report at the offices of the South Downs Society in Pulborough attended by myself, Chris Todd, Ruth Chambers and Margaret Paren to discuss what to do. There was little need for discussion: we would fight on!

CHAPTER FOURTEEN

THE FIGHT BACK

For a long time, the activities of the South Downs Campaign had been relatively low key, but the arrival of the Inspector's report in 2007 galvanized us into a frenzy of activity.

We wanted to see the case for the market towns to be included, most notably Lewes. We wanted also to support the inclusion of the delightful village of Ditchling and more. But the most serious omission, in terms of land area, was the Western Weald, stretching from the east of Petersfield to Petworth, encompassing an area larger than the Broads National Park.

We recognised that to overturn the findings of the Inspector's report there would need to be a new or re-opened Inquiry that would take evidence on the Western Weald and the other omissions. However, Defra planned a short consultation of six weeks from 2 July to 13 August 2007 which was limited to matters arising from the amendments in the NERC Act; the Meyrick judgment; the possible new boundary drawn up for Natural England based on the Inspector's findings; and additional areas recommended by the Inspector.

We were deeply concerned about the narrow scope of the consultation. There was a hint in the letter that went out to all who had given evidence to the first phase of the Inquiry that it might be re-opened, although we heard that Defra planned a short Inquiry of no more than a week. This would not do. We were determined to campaign for the Inquiry to be re-opened and to allow the case for the omitted areas to be made in full.

The immediate priority was to get an extension to the consultation period. By running a six-week consultation (and in the summer holiday period) Defra were not following Cabinet Office best practice guidance that called for a consultation period of twelve weeks.

Fortuitously, we had previously decided to keep the Campaign in the public eye by arranging a celebration of the 60th anniversary of the Hobhouse Report. It was to be a walk organised by Owen Plunkett of the Hampshire Ramblers from Harting Down, a stunning location on the South Downs Way with immense panoramic views. From here, almost the whole sweep of the Western Weald is laid out before you. It is a scene that many a visitor, from other parts of the country or overseas, regards, and then says: 'This is England'.

The day chosen for the walk was 8 July. It was a perfect, sunny day, and the numbers attending, no doubt swollen by the news of just 6 days earlier, totalled over 100. Banners had been hastily produced reading 'Save the Western Weald' or demanding the inclusion of a particular town or village. We were joined by the larger-than-life actor Brian Blessed, the outgoing president of CNP. His presence ensured strong media attendance and a loud and erudite message.

Brian Blessed on Harting Down
© *South Downs Campaign*

Meanwhile, 20,000 flyers were being produced and distributed. Letters were sent to the new National Parks Minister, Jonathan Shaw, expressing concern at the consultation process and the exclusion of the Western Weald; and to Natural England urging them to stick by the Designation Order boundary. We were advised that the Natural England Board were prepared to defend the Designation Order boundary, which was a great relief given the Countryside Agency's earlier failure to do so in the case of the New Forest.

The opposition to the Inspector's proposal to exclude the Western Weald from the National Park was not confined to members of the Campaign. A huge number of local residents, including some who had originally been against the

National Park, were appalled at the prospect of being excluded. They were nervous that this could lead to the loss of protected status as an AONB and angry that their countryside had somehow been found wanting.

Members of the Executive and others were active in engaging with town and parish councils. Christopher Napier, who not only chaired CPRE Hampshire but was also a National CPRE Trustee and had previously been a senior partner in an international law firm recalls:

'I well remember preparing and delivering, with Margaret Paren on behalf of CPRE Hampshire, national CPRE and the South Downs Campaign, what we considered to be a crucial presentation at a specially convened seminar on 26 July 2007 for town and parish councillors and representatives of interested organisations in the council chamber at Penns Place, Petersfield, on the need for the Western Weald to be included within the new South Downs National Park in the national, regional and local interest. I emphasised that the Inspector was wrong to leave out the Western Weald on the basis that it has a different 'characteristic natural beauty' to the chalk downland, a view for which there was no legal foundation. How the two areas had always been treated together as 'the South Downs', with strong visual and historical integration, and common management within the two AONBs; how, if confirmed, the management focus in terms of funding, staff and rangers would be bound to shift to the National Park, leaving the Western Weald as a second class citizen even if it retained AONB status; and how the Rother Valley and A3 corridor had hardly changed at all since AONB designation in the 1960s.

This seminar was immediately followed by an emergency meeting of East Hampshire District Council in which many of the councillors will have attended the seminar and heard our presentation. To our relief, this presentation seemed to find its mark, as councillors voted by a significant majority to write to the Secretary of State to support inclusion of the Western Weald, including Petersfield, Liss and the A3 corridor. The Leader, Elizabeth Cartwright, summed up the general feeling of the meeting when she said "We owe it to our residents to send out a very strong message about large tracts of our area being excluded." Hampshire County Council also came out in support.'

Two Hampshire Conservative MPs also gave support. James Arbuthnot wrote about his concerns to the Minister and Michael Mates fought for Petersfield and Liss to be included. Len Clark reported that he had received a reply from his local Conservative MP, Jeremy Hunt, who welcomed the National Park but was concerned at the exclusion of the Western Weald.

I spent a great deal of time on the telephone seeking out members of all the parish councils within the Western Weald in Sussex. This was very productive. Seven of the parishes became members of the Campaign. Midhurst Town Council, led by Colin Hughes, came on board very promptly and was a strong advocate, together with several other Midhurst organisations. John Carden, the grandson of Sir Herbert Carden and a great supporter of the Campaign, rallied the support of the Brighton Labour MPs who wrote to the Secretary of State. Members of the House of Lords from the Brighton area were lobbied too.

I wrote a letter to every Chichester District councillor and spoke to a few of them. However, their council voted for a narrower boundary excluding even more of the Western Weald. Andrew Shaxson, an independent councillor from Harting, expressed his disappointment in the press. *'It was a "poor and shallow" debate. Many councillors had left the meeting before the issue was discussed. There is no guarantee that AONB status will remain in areas outside the National Park. But I am afraid that the councillors buried their heads in the sand.'*

We received a copy of the advice West Sussex County Council had received from their legal advisers on this issue of continued protection should the Western Weald be excluded from the South Downs National Park. We were very concerned by the subsequent statement by the Leader of the council, Henry Smith, which implied that all of the Western Weald would continue to keep its protected status, which was not what the advice had said. We issued a press statement accordingly.

Despite the negativity of these two councils, on 27 July the Minister reacted positively to the pressure and announced that although 'the formal deadline remained 13 August, the Secretary of State (by now Hilary Benn) would consider all objections and representations which were received by 24 September'.

The extra time now available was put to good use. The public campaign continued to gain momentum. John Venning, who had initially opposed the National Park when an East Hampshire District councillor, was by now chairman of the East Hampshire District of Hampshire CPRE and an enthusiastic supporter of the Campaign. He took on the task of organising a written petition to Parliament. Christopher Napier remembers:

'Speaking and participation at meetings of amenity societies and parish councils and canvassing of the public in the street and outside supermarkets became a way of life over the weeks of the consultation period. Overwhelming support was expressed to us for inclusion of the Western Weald, with Petersfield and Liss. Many signatures were collected on a written petition, including over

1000 in Petersfield on the August bank holiday weekend, and a government online petition set up by the Campaign was well supported. Michael Mates, MP for East Hampshire, publicly signed the written petition.'

The local papers in both Hampshire and West Sussex were full of letters from people objecting to the exclusion of their local areas. Many criticized West Sussex County Council and Chichester District Council for their lack of support. Perhaps the most eye-catching letter was one in the Midhurst and Petworth Observer from 15-year-old Sophie-May Lewis which included a poem:

'*I have written this poem to encourage people to look around them while walking on the South Downs Way, in the hope they may appreciate the surrounding area and not just see the chalk ridge upon which they tread, beautiful though it may be.*

The Western Weald

If you have stood upon the Downs,
Whilst flowers dance about your feet,
With butterflies with dainty wings.
If you have watched a soaring speck,
And listened to the song,
Wondering how it stays aloft that long?
If you've ever seen a Red Kite glide,
On the thermals where the Buzzards hang,
Watching a hare, as through the grass it ran,
You may never have travelled down the hills,
To the valley that lies below,
Here you can wander through village streets
Where colourful bunting blows, and
Children's laughter echoes, across the green,
Where you can feel the scented south-west wind,
From across the flowering Lowland Heath,
And honey bees hum to the warmth of sand,
Will you stop to stand and stare?
And wonder, what would life be like?
If these places, were no longer there?
All could be, so easily lost,
If we don't stand up, and show we care.'

In our determination to raise the profile of our campaign for the Western Weald we persuaded several national figures to lend their support. The actress Honeysuckle Weeks made a strong public statement and David Dimbleby, who had previously chaired a special meeting for us in Eastbourne, spoke up for us once again.

This culminated in a day-long visit on 11 September 2007 by Bill Bryson, then President of CPRE. The visit was managed within CPRE by one of the newer members of our Executive, Emma Marrington. Her love of the countryside had been inspired by youth hostelling trips with her mother who, it was rumoured, had walked the Seven Sisters when pregnant with Emma. In Emma's own words:

'With my family background it was wonderful to get involved with the Campaign. Not only was the cause so aligned with me personally, I loved getting to see so much of the landscape both on the way to meetings and scoping missions to get to know it better. I also developed "super hearing" for when the tea trolley was coming along the train to and from meetings!'

The day chosen for Bill Bryson's visit was bright and sunny. He spent the morning in Hampshire and the afternoon in West Sussex. The day began at Ditcham Barn, a stunning tithe barn clad with brick and flint, nestling under the scarp slope of the Downs on the border of Hampshire and West Sussex. Here he was briefed by John Venning, Chris Todd and Margaret Paren. By late morning they had reached Longmoor, the tract of exceptional heathland forming part of the Ministry of Defence estate, despite which it had almost continuous public access. They climbed a hillock from where there were views down the wooded Upper Rother Valley to Butser Hill, the highest point on the South Downs ridge. To the west lay the Hampshire Hangers and to the east Hillbrow: a ridge that marks the boundary between Hampshire and West Sussex. Margaret Paren recalled that Bill Bryson took in the scene and announced he did not need to see more. He added that did not mean that he didn't want to see more, just that he was convinced of the case.

After lunching with a group of us at Elsted near Midhurst, Bill Bryson visited parts of the Western Weald in West Sussex before finishing the day by attending the first part of Campaign's executive meeting in Chichester. As intended, the visit brought extensive national and local media coverage, with photographs of the threatened area. Signing the written petition, he urged the Secretary of State to stick to the Countryside Agency boundary; and standing at a point within the Western Weald from where he could see the majesty of the chalk downland, he summed up the situation well when he said:

> *'This countryside is some of England's finest and one of the glories of the English landscape is its huge variety. The land I am standing in is gorgeous and the land I am looking at is gorgeous, yet one is in the National Park and the other is not. It's bizarre.'*

Following the visit, Bill Bryson wrote a two-page article for the Independent on Sunday supporting the Campaign and questioning the rationale for the Inspector's decision. It later became clear the Inspector had read the article and rather hoped Bill Bryson would attend a hearing before him.

We realised that we had to take the lead in defending the Western Weald, covering as it did so much countryside and so many towns and villages. But we recognised that would stretch our resources to the full. We therefore reluctantly took the decision that in the case of other exclusions, it would be up to the local communities to take the lead and if they did so we would provide advice and support.

The most important of these other exclusions was Lewes. It had been omitted from the Sussex Downs AONB but it, and the Ouse flood-plain to the north, had been included in the designated boundary. In his report, the Inspector recommended exclusion. His reasoning seemed to centre around the fact that Lewes had not been within the AONB, and he doubted that the area to the north of the town met the designation criteria. He also said that if Lewes were to be included it would be a very large settlement by national park standards and from vantage points within the surrounding countryside it appeared as a significant urban intrusion into an otherwise pastoral scene.

This rejection was met with alarm by those who had fought for the town's inclusion. In particular, the civic society, the Friends of Lewes, considered that concerted action needed to be taken to attempt to reverse the Inspector's recommendation. This would mean persuading the town's residents, the Town, District and County councils as well as local MPs, that a national park would be of long- term benefit to the town and to say so.

The Chairman of the Friends of Lewes at that time was Robert Cheesman. He had joined our Executive in 2002 and was one of our many erudite retired Civil Servants. Born and bred in Lewes, he had lived there most of his life save for a few years when his work took him to Cumbria and, like many of our Executive members, he was an active volunteer in a number of societies. Amongst these, most notably, was perhaps his role as County Commissioner and then County Chairman of the Scout Movement. Later he was to become chairman of the South Downs Society.

One of the first moves was to hold a public meeting, which was chaired by Lord Renton, about the advantages that national park status could bring to Lewes. The main speaker in favour was Jack Ellerby who at the time was a planning officer for the Exmoor National Park Authority and he enthused the meeting with his positive thinking, in particular, that a national park would have a permanent authority that would protect the town and countryside from unwelcome development. The strong relationships that the Friends of Lewes had with the Town and District councils persuaded these two local authorities to support the case. Sadly, East Sussex County Council, with whom a strong relationship did not exist, was unpersuaded. It was interesting to learn subsequently that Steve Ankers, their witness at the Inquiry, did not agree with the decision his political masters had taken.

Working closely with Robert Cheesman was Paul Millmore, who had joined the Friends of Lewes committee. Together they organised a petition that attracted nearly 3,000 signatures. This was presented to the Government in January 2008 by the Town Mayor and Norman Baker MP. Although the latter had for some time been unenthusiastic about the National Park, he had by this time come down in favour of it when he could see the extent to which the public in the town supported its creation. All the while, Paul worked frantically to dig up relevant historical and cultural relationships between the town and Downs that could be used to make the case for Lewes.

By contrast, we were disappointed by the reaction in Arundel. Positioned as it is above the River Arun and its floodplain, with its magnificent castle and cathedral, it is one of the most iconic towns in England. To argue for its inclusion after the strong case made against it by the witnesses for West Sussex County Council was always going to be an uphill struggle. But added to that, the town was divided between those who wanted to see it included and those who feared its inclusion might jeopardise the case for a new by-pass. We therefore decided with deep regret not to pursue the case for Arundel. However, we didn't abandon all hope. We worked with local people to push for more of the landscape to the west of Arundel to be included.

We had also been disappointed that Steyning, had not been included but there seemed little sign of a local campaign at any stage to rectify this. Apparently there had been a discussion in the village but the attitude of West Sussex County Council and the local MP, Tim Loughton, had persuaded them not to pursue their case.

But the residents of Ditchling were not about to take their exclusion lying down. The well organised Ditchling Society, together with the Parish Council, set about gathering a wealth of evidence about Ditchling's rich culture and strong downland connections. This built upon the already substantial evidence that the Society had collated at the start of the Inquiry. Chris Todd was very involved in advising them and helping to draw up the case for the inclusion of the settlement and land to the north, necessary in order for Ditchling itself to be considered.

Whilst the public campaigns flourished, we began the task of putting together our consultation response. We had become deeply concerned about the Landscape Assessor's reports on which the Inspector had apparently relied in reaching his conclusion that the Western Weald did not merit inclusion in the National Park. His report on the area, which he claimed no longer met the natural beauty criteria, drew on field work he had undertaken almost certainly in his own time; and the examples provided in his report suggested a narrow concentration on the A3, the A272 and a short stretch of the Sussex Border path. Our Freedom of Information requests to the Planning Inspectorate and Scottish Heritage bore fruit. We were shocked to find that the report had not only been delivered after the public hearings but after the formal closure of the Inquiry, thereby making it impossible to challenge the findings other than by a re-opening of the Inquiry, which was still not certain.

On 24 September 2007 we responded to the consultation and separately wrote a letter to the Secretary of State at Defra, Hilary Benn. In this letter we enunciated our concerns about the process and the veracity of the Inspector's findings, particularly given the amendments in the NERC Act and the outcome of the Meyrick case. We conveyed our concerns regarding the proposition that the National Park should comprise a single character type, and that the Inspector had apparently been influenced by non-statutory designation considerations such as planning. But the main thrust of our letter was about the exclusion of the Western Weald. We drew the Secretary of States' attention to the Inspector's reliance on the Landscape Assessor's report, delivered after the closure of the Inquiry, which questioned the quality of the 'A3 corridor' and the Lower Rother Valley, and could lead to the possibility of the largest ever de-designation of any nationally designated landscape. We questioned why the positive case for the Western Weald had not been heard at the Inquiry. We suggested that in these circumstances any decision based on the Inspector's report could be open to legal challenge.

We were not the only ones pressing our case on Defra. We became aware that an officer at West Sussex County Council had written to Defra officials at the beginning of August:

'I understand that the South Downs Campaign has been lobbying to have any second Inquiry reconsider the issue of whether the Western Weald should be included in the National Park. In my view Defra has wisely decided to exclude the principle of including the Weald from any second Inquiry. I would be grateful therefore for your confirmation that Defra are not considering extending the scope of any second Inquiry to allow the principle of the inclusion of the Weald to be considered.'

As the closing date for the consultation passed, we continued to campaign as we awaited the response from Defra. The TV presenter and adventurer Ben Fogle, who was the new President of CNP, visited us on 10 October together with Brian Blessed and Kathy Moore, CNP's new Director. Chris Todd, Margaret Paren, Richard Reed and I took them to Older Hill, owned by the National Trust. Here there are breath-taking views over a carpet of mainly ancient woodland looking towards the South Downs ridge and the Hampshire Hangers, with another stunning view northwards towards steep greensand ridges and Black Down.

This time the wait was not too long. On 25 October 2007 Jonathan Shaw announced in a letter the re-opening of the Inquiry. In it, he repeated the scope previously set out in the consultation but, crucially, asked the Inspector to indicate any other points raised during the consultation that would cause him to change his mind on the findings in his first report. We could hardly believe our eyes: the fight was on!

CHAPTER FIFTEEN

THE INQUIRY RE-OPENS

The Minister's letter stated that the pre-Inquiry meeting would take place on 11 December 2007 and the Inquiry itself would re-open on 12 February 2008.

A lot of work needed to be done in a short period of time. The first major hurdle was the pre-Inquiry meeting, which was to be held at Worthing Town Hall. This would be crucial in persuading the Inspector to hear the case for the excluded areas. We heard that Natural England were planning to brief Robert Griffiths again and that Hampshire County Council had hired a barrister to make their case for hearings on the Western Weald.

We decided the event needed to be high profile to help convince the Inspector of the popular support for the omitted areas and to extend the remit of the next phase of the Inquiry. We therefore encouraged local groups to travel to Worthing and banners were put together in support of the towns, villages and countryside at stake. A particularly strong contingent arrived from Hampshire, supported by all three prospective Parliamentary candidates for East Hampshire, including their current MP, Damien Hinds. With around 60 in the gathering, we heard later that the Inspector had asked his administrative officer to photograph the event for his records.

The purpose of the pre-Inquiry meeting was to prepare for the actual Inquiry by discussing procedural and other arrangements. Much the most important issue was how the Inspector would deal with the added leeway he had been given by the Minister. Legal representations were made by both Natural England and Hampshire County Council backing the need for hearings on the Western Weald, and they were supported by Ruth Chambers on our behalf. Christopher Napier recalled a well-attended meeting, which became quite agitated at times. A threat of judicial review of any boundary which did not include the Western Weald was clearly voiced. The Inspector said that he would accept written representations from anyone who opposed the original decision, including the exclusion of the Western Weald, but would only make up his mind on the first day of the Inquiry whether to allow actual discussion of the matter. This was better than nothing, but still unclear. Even if he allowed discussion, we recognised it would still be

A message to the Inspector from the Campaign
© *Chris Todd*

The inimitable Paul Millmore with Robert Cheesman
© *Chris Todd*

an uphill struggle to get the Inspector to change his earlier view, publicly stated in his report. This left all of us in suspense and not a little anxious.

Meanwhile, our committee decided that, for this second phase of the Inquiry, we needed professional help to put our case for the Western Weald and the market towns. With aid from Hampshire County Council a specification was drawn up and sent to some of the leading consultancies in the field. The sum of money we had available, which I and others had worked hard to procure, was substantial by our standards but modest in the marketplace. However, we received an expression of interest from Professor Robert Tregay, who had drawn up the original boundary. Understandably, being dropped from the first phase of the Inquiry by the Countryside Agency still rankled with him, and he saw our approach as a means of reconnecting with the process.

We held our first meeting with him early in December 2007 and were much encouraged by what he had to say. He thought that, although the Landscape Assessor's reports were well-written, he had concluded that they were fundamentally flawed, and a robust and persuasive argument could be put to the Inspector. He also gave us confidence that the key arguments against the omitted areas could be overturned. Nevertheless, he was hesitant to work with us, being concerned that an organisation made up of so many disparate voluntary groups might be unwieldy and lack clarity in its aims. We managed to persuade him that we had a robust procedure for determining what we were prepared to endorse, and he would receive clear direction on what we wanted from him. We therefore engaged him to help prepare for the Inquiry and appear for us if allowed to do so by the Inspector. We also promised him whatever photographs he needed and commissioned Richard Reed to set to work as necessary.

By this stage Margaret Paren had become the de facto leader of the Western Weald campaign and John Venning her de facto deputy. Margaret recalls a delightful man who would ring up most mornings to ask what he could do to help. The team they put together and inspired performed wonders in the little time that was available to assemble our evidence. Of our meeting with Robert Tregay she recalled two insights that became key in her planning for the re-opened Inquiry:

'The first observation was that many non-professional Inquiry witnesses fell into the trap of including too much superficial detail in their evidence and thereby left inspectors confused as to what were the critical matters. Later, based on this, I drew up a tight specification for all the Proofs of Evidence that were produced to ensure a clarity of message within and between them, so the

Inspector remained focussed on the key issues.

The second was with regard to the towns and the planning load they represented for the National Park Authority. This we had wrestled with ourselves in the past and rather retreated into defensiveness. Robert made me realise that having the towns in the National Park was crucial as it meant the National Park Authority could plan their future as an integral hub or gateway without recourse to a separate planning authority.'

Following the pre-Inquiry meeting, we met with Natural England and Hampshire County Council to determine how we would organise the evidence to be put to the Inquiry. This was a perfectly proper approach as it prevented the Inspector having to hear substantially the same evidence from separate sources and reduced the time that needed to be devoted to the Inquiry. This meeting brought us face-to face with two people that were to become important players in the months ahead. First was Linda Tagtallia-Kershaw, who led the award-winning landscape team at Hampshire County Council. The second was the Natural England officer tasked with supervising their involvement in the Inquiry, Chris Fairbrother. He proved to be a great asset: always calm, at least on the surface, and managing a limited budget to the greatest possible effect.

The meeting decided that Natural England should concentrate on the principle and policy issues; Hampshire County Council on qualitative and quantitative information; and the South Downs Campaign on public support, and a detailed rebuttal of the findings in the reports from the first phase of the Inquiry.

An extraordinary amount of work was put in over the period December to March 2008. Under the leadership of Margaret Paren and the co-ordination of Chris Todd we produced eighty-five documents by 23 authors from a variety of organisations. Margaret recalls this critical period in the Campaign history:

'Getting together the evidence, particularly that necessary to refute the many unsubstantiated assertions by the Inspector's Landscape Assessor, was a monumental task and only achievable because of the breadth and depth of knowledge to be found amongst the communities and many and varied local and regional societies. Nor was there a lack of volunteers to turn the welter of evidence into the necessary documents. The problem, as Robert Tregay had foreseen, was to curb the natural enthusiasm of participants and to ensure they kept to the specification provided. As the drafts became available, they were rigorously peer reviewed, and any unsubstantiated evidence got short shrift. Consequently, we were able to go into the second part of the Inquiry with confidence that the evidence was not only robust but amounted to a strong and

coherent case for the Western Weald remaining in the National Park.

It was inspirational to see the enthusiasm and dedication of so many people in the cause.'

Once the proofs and their fat appendices containing the evidence had been completed the task of assembling and binding six copies of each was undertaken. This was more than Chris Todd could cope with given everything else he was doing at the time, so Jacquetta Fewster, the South Downs Society's recently recruited Director, and the deputy clerk at Liss Parish Council took on the task.

We were conscious that the Inspector would be paying site visits to the disputed areas. Chris Todd undertook to produce posters, all on green paper, which read 'Keep (place name) in the National Park'. These were distributed to townspeople and villagers who fly-posted their areas so the Inspector would get a clear message about their views.

© *Chris Todd*
Andrea Mann, Jacquetta Fewster and Margaret Paren assembling a wealth of the Campaign's evidence

Whilst this hard work was underway, pressure was mounting on West Sussex County Council from the parishes in the Western Weald. I attended the meeting with the Leader of the County Council, Henry Smith on 31 January 2008, organised by a now dissenting County Councillor, Nicola Hendon, who had stated that fears over loss of AONB status were 99% real. She brought with her parish and town council representatives from Harting, Elsted, Linchmere, Midhurst, Rogate, Milland, Stedham with Iping, Woolbeding, Tillington, Fernhurst, Northchapel, Bepton and Fittleworth. Councillor Henry Smith repeated his *'cast iron guarantee'* that the county would *'fight tooth and nail'* for retention of AONB status for the area left out of the National Park.

Two weeks later West Sussex County Council issued a press statement which read:

'There continues to be no support for inclusion of the Wealden area in the South Downs National Park.'

We awaited the first day of the re-opened Inquiry with nervous anticipation. We believed that the evidence we had accumulated could, if necessary, be manipulated into the headings identified by the Inspector at the pre-Inquiry meeting for oral hearings. But it would have been fragmented and our strong narrative would have been difficult to maintain.

The Inquiry once more took up residence in the Chatsworth Hotel. By now the media was following the story with great interest. Margaret Paren and Phil Belden had appeared on Countryfile the previous week and Margaret and Chris Todd had to miss the first minutes of the Inquiry to give interviews which appeared on regional television news programmes that evening.

Ruth Chambers summed up the first day in an e-mail to Chris Todd, Margaret Paren and me:

'The good news is that the Inspector will allow oral submissions on the Western Weald and that these will take place in separate, additional sessions. This is a joint success for Hampshire County Council and the Campaign (Natural England had been somewhat ambivalent on this) but with lots of support for this from the parishes. The Inspector said it 'would be a nonsense for the Secretary of State not to take account of fresh and current material', recognising that it had been some time since the Inquiry last sat. The Inspector will hear "what people want to say on how the Assessor's report and Inspector's conclusions were drawn up", without going so far as to make himself available for cross-examination!

West Sussex County Council's position is becoming more interesting – its Counsel confirmed that it had no 'in principle' objection to the hearing of new evidence on the exclusion of the Western Weald and said that in its view that decision was entirely down to the Inspector. WSCC also accepted that the Inspector would find it difficult on the grounds of natural justice and the need to report fully to the Secretary of State not to hear new evidence on the Western Weald. This removal of any final barrier to a separate session on the Western Weald made it easier for the Inspector, but he suggested he had already made his mind up on this before the start of the Inquiry. Finally, WSCC said it supported a 'fair and balanced' decision which prompted a few boos and much ironic laughter – even the Inspector was smiling at this point.

WSCC doesn't yet know whether it would wish to appear in any sessions on the Western Weald. The reason for this shift in position is the immense and unrelenting pressure it has been under, more latterly from many angry parish councils.

All in all, a very good start!'

There followed evidence on the legal implications of the Meyrick Case and amendments to the NERC Act. Hampshire County Council gave their evidence first followed by Natural England and then Christopher Napier for the Campaign. During these sessions it was noticeable that the Inspector was quite tense. We were concerned: did this mean he was antipathetic to the case? Strong representations were made that the changes in the NERC Act clarified that single core character was not relevant.

We used this part of the Inquiry programme to make our case for the inclusion of Lewes and Ditchling. Evidence on Lewes was given by Robert Cheesman and Paul Millmore. In essence their case was that the NERC Act amendments had clarified that wildlife and cultural heritage were factors to be taken into account in the designation process, that the Ouse Valley had wildlife in abundance and Lewes a wealth of cultural heritage, so both should be included. In his report the Inspector had said: '*I am not convinced that Lewes can be said to be deeply embedded in a landscape of national park quality.*' To counter this, when Paul Millmore gave his evidence, he memorably produced a large ball of clay and plunged a large and lethal-looking knife into it, saying that, by being surrounded on at least three sides by landscapes that met the natural beauty criterion, the town must 'de facto' be deeply embedded. The Inspector looked duly impressed.

The Inquiry then moved on to the eagerly awaited sessions on the Western Weald. A series of Parish Councils and other local groups made their case. We heard that the Inspector was enjoying himself and looked much more relaxed. During these sessions, much to our amazement, he had said that the indicative boundary he had included in his first report was not meant to be as precise as it looked, it was meant only to demonstrate in broad terms where it might lie.

There had been a good deal of nervousness as to how the Inspector would take criticism of his earlier findings. Natural England, Hampshire County Council and ourselves each concluded that the real villains of the piece were the two reports by the Landscape Assessor, and these had to be attacked: no other option existed if we were to win the case. But we were uneasy about how the Inspector would take this, given the time they had spent together and the reliance he had placed on the reports in question. In these earlier sessions, the legal teams from Natural England and Hampshire County Council at first tentatively and then with growing confidence, dissected the reports' findings with little or no signs of concern from the Inspector.

Our hearings were scheduled towards the end of the Inquiry and Chris Todd had negotiated with the Inspector's administrative officer that we should have three full days in one week to give our evidence followed by a day the following week for Robert Tregay. But before we reached this point, we observed progress as others made their case.

Hampshire County Council's witness based his evidence on the South Downs Integrated Landscape Character Assessment of December 2005, painstakingly comparing the area in Hampshire excluded in the Inspector's report, the so-called 'A3 corridor' and 'Bentley Nib', with that which had been included by the Inspector in the 'chalk-only' national park. He concluded that the cultural and semi-natural characteristics were of a high quality, often superior to those within the Inspector's boundary. He also demonstrated that the excluded area provided potentially superior recreational opportunities to the included area.

As we prepared for our own appearance, we agonised over the tone to adopt in our evidence, particularly our opening statement that would set the scene for the following sessions. The statement, a careful mix of fact and passion, was delivered by Margaret Paren and followed by a slide show to provide the Inspector with a flavour of the beauty of the area in question. We made three main points. First, we were supporting the designated boundary and making no case at this stage for additions to the area previously put to the Inquiry but rejected. Secondly, we said that contrary to the Landscape Assessor's assertions, public opinion had strongly favoured the inclusion of the Western Weald throughout the process of designation. Thirdly, that the Landscape Assessor had claimed that an area equivalent in size to a quarter of England's AONBs had become so degraded it no longer met the natural beauty criteria. If true,' *there had been a profound breakdown of national, regional and local policies directed not just at the conservation of natural beauty but its enhancement'.*

We then entered upon a series of sessions each linked to a key matter that had led the Landscape Assessor to his conclusions. First off was the issue of intensification of farming in the Lower Rother Valley which the Landscape Assessor claimed had led to field enlargement and the loss of hedgerows. We had accumulated evidence which demonstrated that there had been very limited enlargement of fields, that aerial reconnaissance photographs from World War 2 showed field boundaries had been varied then as now and that hedgerows had been planted rather than destroyed. Our witnesses demonstrated that farming had not degraded this part of the landscape since the AONB had been established in 1966 and, thanks to environmental stewardship schemes, that significant landscape enhancement had been undertaken more recently.

In the afternoon we moved on to AONB status if the Western Weald was outside the National Park, an issue addressed by all the main players at some stage in the Inquiry.

The following morning was a round-table session on the major settlements and so-called landscape detractors in the Rother Valley. Evidence on the settlements in West Sussex was given by Emma Marrington from national CPRE who debunked the suggestion that the settlements had suffered from intrusive development that had diminished the quality of the surrounding countryside. Rodney Chambers, the Chair of Sussex CPRE, concentrated on so-called landscape detractors. Thanks to Andrew Shaxson's good memory and Richard Reed's delving through the South Downs Society's archive, we were able to show that the power line which figured significantly in the Landscape Assessor's report, had been planned and largely constructed before the Sussex Downs AONB was confirmed. And, with statistics from West Sussex County Council, that the A272 was no noisier than other comparable roads nor noisier through the Weald than the chalk.

Tony Struthers of the Petersfield Society gave evidence on Petersfield and was also able, as a retired Planning Director, to add perspective to the history of development in the area. He and Nigel Paren, speaking on behalf of Liss Parish Council which he chaired, were able to prove that all the significant development in Liss and Petersfield had been planned before the AONB came into being and that neither settlement impinged on the wider countryside. Based on his own research, sitting for days by the railway track timing the period when the trains could be heard as they passed through Liss, Nigel Paren was able to refute any suggestion that noise from the railway impinged on the quiet enjoyment of the area.

This afternoon session was devoted to the A3 itself and its alleged impact on the landscape. Cllr Sue Halstead, drew on proofs of evidence from the inquiries in 1985 and 1988 into the new road, which she had retained in her records. She demonstrated that the AONB status of the area had been taken into account in the landscaping of the road and, as cabinet member for planning at East Hampshire District Council, she was able to state unequivocally that the road corridor had subsequently been protected from development.

Margaret Paren brought to a close the day's session with evidence on Woolmer Forest and Longmoor. These Ministry of Defence owned heathland areas had been excluded from the AONB but subsequent change in usage had dramatically improved their landscapes, and their wildlife value was now internationally significant. The Landscape Assessor had not attended the relevant session in the first part of the Inquiry and relied, instead, on the written evidence provided by the Ministry of Defence. Having been involved many years earlier in the production of the Ministry's evidence for the Stansted Inquiry, Margaret Paren was appalled by the quality of what the Ministry presented to this Inquiry. Her evidence easily refuted what the Ministry had said. A survey had been conducted on users of Longmoor. Another survey debunked the Ministry's evidence on the frequency of Woolmer being closed to the public. The history of the area was researched from books and photographs owned by local people as well as information provided by the retired colonel who had run the area. No less than seven local groups from Petersfield, Liss and Whitehill, together with Hampshire CPRE, put together a written representation on Longmoor and Woolmer Forest in support of the main proof of evidence.

By the third day the Inspector seemed positively relaxed. We started the morning with a session on recreation by Jacquetta Fewster, Brian Cheater a retired Royal Air Force Officer and keen rambler, and Ben Perkins. Their evidence too had been thoroughly researched: Jacquetta provided the results of a survey of over 100 walking, cycling and horse-riding groups about the quality of recreation experience in the Western Weald and compared it favourably with those available in the Surrey Hills AONB.

It had been Brian's task to complete all fourteen walks listed in the AONB publication 'Rother Valley Walks'. Starting in Petersfield and ending in Pulborough it required him to walk 95 miles, though he walked many more because of the many attractions along the route. At the hearing he could therefore confidently refute the

Landscape Assessor's assertions in his report that '*...many paths are overgrown and difficult to follow*' and '*...closer to the River Rother the footpaths are fewer and fragmented*' and finally that '*the bridges tend to be traffic bottlenecks, making it difficult for recreational users, particularly at peak times.*' In short, he found the Landscape Assessor's report to the Inspector had been superficial and incomplete. He recalled being asked how he rated the Western Weald against trekking the Italian Dolomites or Nepal and was able to say that it gave him the same spirit lifting experience that he had enjoyed whilst walking in other very special places.

Scheduled in for the afternoon was the proposed new boundary drawn up by Natural England, which was based on the Inspector's indicative boundary. We had anticipated this would be a crucial session. A great deal of effort had been put in to prepare this evidence including virtually the whole of it being walked by volunteers: Roger Mullenger, the Vice Chair of Liss Parish Council the area within that parish, John Venning around Petersfield and Steep, Andrew Shaxson from the Hampshire border to Elsted, Jacquetta Fewster around Midhurst and Rodney Chambers and others from Sussex CPRE the area eastwards. Our witnesses, Margaret Paren and John Venning, had prepared carefully. The aim was first to demonstrate that the proposed boundary was not fit for purpose. That if the 'A3 corridor' was to be excluded there was no boundary on the table. Then we contended that as the only other boundary that had been subject to public scrutiny was the designated boundary, if the Inspector were to recommend any other boundary there would need to be another consultation and Inquiry. Margaret Paren recollects:

'*When we got to the session, the Inspector re-iterated what he had said earlier, that his indicative line had been taken too literally, which rather reduced the scope of the session. We got into a conversation about whether, in defining a boundary, you should rely on a line on a map, a description, or both. John Venning then told a story from his days in the Foreign Office about the Line of Control in Kashmir. As the Inspector seemed interested, I added a story I had heard from the Nigerian ex-Minister for the Interior, about the problems created by the colonial boundary between Nigeria and Benin. Not to be outdone, Robert Griffiths threw in a story about boundary problems in Malaysia. All in all, an entertaining and relaxing interlude!*'

Our trump card was Robert Tregay, to appear for us as an expert witness. A particularly telling part of his evidence was:

'*My original recommendations to the Countryside Agency on the broad extent of the National Park, including the specific inclusion of the Western*

Weald, were entirely based on my own professional analysis and judgement. I was not asked by the Agency, or anyone else, to support any particular extent or type of national park. The matter was open from the outset. I had no axes to grind or political pressures to bow to. Whilst consulting very widely throughout the boundary-setting process and receiving advice from statutory and non-statutory stakeholders and consultees, my professional recommendations were always independent and based on thorough analysis of the legal and policy basis for national parks and of the South Downs itself. Having reviewed my original conclusions on the inclusion of the Western Weald within the proposed National Park, I see no reason to change my original recommendations and, in fact, the additional evidence in this proof adds further weight to the case for the Western Weald to be included in the proposed National Park.'

Robert Tregay took the Inspector through the process that identified the Area of Search and then the draft boundary and in doing so he refuted the criticisms mounted against it by the Inspector's Landscape Assessor. He summarised the reasons for including each part of the Western Weald, drawing on new evidence, including that we had provided the week before, to support his proposition that the original decision to include the whole of the Western Weald was the right one. He was also highly critical of the paucity of methodology employed by the Landscape Assessor in reaching his conclusions and that any decision based on his reports would be unsound.

This day-long *tour de force* was listened to in silence: even Robert Griffiths failed to intervene. There was no doubt in our minds that we had made the right decision: if nothing else, the Inspector was able to judge the man for himself rather than through the prism of those who wished to refute his conclusions. He could not fail to have been impressed.

To this point in the proceedings there had been one notable absentee: Moira Hankinson on behalf of the councils. She had an extremely difficult task ahead of her. On the one hand she wanted to support the Inspector and his Landscape Assessor in their conclusion that the National Park should exclude the Western Weald. However, given the reassurances the Leader of West Sussex County Council had given to communities on the continuation of protected status for the area, she could not support the Landscape Assessor's contention that part of the area had lost natural beauty. Moreover, in line with the decision of the councils to support a boundary closer to the chalk, she had produced yet another proposed boundary in response to the public consultation, albeit one with alternatives at several junctures and which stopped abruptly at the Hampshire border.

By this stage of the Inquiry, the legal teams for Natural England and Hampshire County Council had grown weary of attacking the Landscape Assessor's reports, there was little more that could be said about them. So the appearance of Moira Hankinson afforded a fresh opportunity to show off their cross examination skills, one they clasped with great alacrity. It was all rather brutal, and one could not but feel sympathy for her. Moira Hankinson's discomfort was apparent as she struggled to fulfil her nigh on impossible remit. The Inspector became visibly annoyed and at one point observed that he had yet to hear any new evidence from the witness.

Alison Farmer, Natural England's landscape expert witness, also had a difficult job to do. She had been commissioned by Natural England to define a detailed alternative boundary based on the Inspector's indicative line and the Landscape Assessor's report. But she had also been commissioned by the same body to defend the original designated boundary. In carrying out this latter task she had revisited the work she and Robert Tregay had carried out during the designation process and applied the now updated methodology. This she had very recently used in her work in the north-west of England on the extensions to the Lake District and Yorkshire Dales National Parks. She was therefore *au fait* with its application to national parks more widely than the South Downs.

In the circumstances it was inevitable that much was made of the fact she had drawn up an alternative boundary she did not support, and she was forced to defend the professionalism of her approach to that exercise.

In summarising her evidence, she re-iterated the point made by Robert Tregay, that there had been no pre-determination that the boundary should be drawn widely to include the Western Weald, it had been included on its own merit. She dismissed the Landscape Assessor's reliance on characteristic natural beauty and described his report as *'misleading and unreliable'*. So far as the areas were concerned that the Landscape Assessor had found to have lost natural beauty, her judgement was that the Rother Valley in Sussex had some limited and localised lower quality areas, but these were 'washed over' by the surrounding high-quality landscape. She regarded the 'A3 corridor' in Hampshire as more balanced, with some pockets of lower quality recreational opportunity, but that the land to the north and east were of high quality. Her closing remarks were:

'In summary, I conclude that the Rother Valley, A3 corridor and the Wealden landscapes all merit inclusion in the South Downs National Park. The proper extent and boundaries of the National Park in these areas should be as shown in the Designation Order.

However, if the Inspector or the Secretary of State, contrary to Natural England's advice, should be minded to exclude any of the areas described above from the National Park, I would draw their attention to the fact that all the land concerned very clearly meets the natural beauty criterion and therefore should remain as AONB land.'

The end of the Inquiry was fast approaching, and it was time to prepare our closing statement along with the other major participants. As Christopher Napier said about the importance of the document:

'The South Downs Campaign's Closing Statement would be crucial in bringing together the evidence given by the Campaign and others in a way which would carry most weight with the Inspector. It was to be hoped that it would be a document he would refer to regularly in writing his Report, so it needed to contain the right mix of evidential fact, conclusions drawn from the evidence, and passion. It was written by Robin Crane, Margaret Paren and me, and ended up at 44 pages. Double spaced, but still quite a lot of work.

On May 28th 2008, we three also read it out in sections, mine the legal section, to the Inspector and a well-attended Inquiry room which was listening very attentively. It went well and afterwards Robert Griffiths QC said to me that it had been 'excellent' and really helpful to the Natural England case.'

In essence we drew the Inspector's attention to the changes since his first report: the impact of the NERC Act amendments and the Meyrick judgments, in particular in relation to the need for an Inspector to explain the reasoning behind his recommendations. We emphasised the substantial body of new evidence: some published, some researched ourselves for the Inquiry. We went on to question the Landscape Assessor's conclusions, that his reliance on single character was now clearly unlawful, that his second report had not been subject to public scrutiny and that his reports contained substantial inaccuracies. Most tellingly, in relation to the Western Weald, his assessment of the loss of quality of the Rother Valley: the A3 corridor in Hampshire and the farmlands in West Sussex, conclusions not shared by Natural England, Hampshire County Council, West Sussex County Council, East Hampshire District Council or Chichester District Council. We then summarised our evidence and concluded by referring to the clear consensus in favour of the inclusion of the Western Weald, Lewes and Ditchling.

Both Hampshire County Council's and Natural England's advocates presented powerful final statements that sought to demolish the arguments presented by our opponents. In his sixty-eight page summing up, Robert

Griffiths began by reminding the Inspector that the confirmation decision had to be made in accordance with the law. The question was whether it was especially desirable to designate the designation order land as a national park, by reason of its natural beauty and the opportunities for open-air recreation, in order to take management measures both for conserving and enhancing the natural beauty, wildlife and cultural heritage of the area, and for promoting opportunities for the understanding and enjoyment of the special qualities of the area.

His second point was that the decision should be made in accordance with the Government's policy on the designation of new national parks, and the more detailed policy made as a result by the Countryside Agency, now Natural England. Having quoted from Michael Meacher's letter to the then Countryside Agency he concluded:

'It is common ground with West Sussex County Council that the Western Weald has outstanding natural beauty, special qualities of national significance in relation to its natural beauty and is a landscape of national importance. It is also common ground with WSCC that the Western Weald does have recreational opportunities. The South Downs National Park, including the Western Weald when considered separately, provide a markedly superior recreational experience when compared with normal countryside, and provides a strong sense of relative wildness.

It is clear that when the words of the 1949 Act as amended are applied to the designated order land, without anything added, or any gloss applied, the extensive tract meets the statutory test for designation. The same applies to the Western Weald if that is assessed separately. The matter is put beyond doubt when the Government's policy on the designation of new national parks is taken into account, and the various considerations weighted as suggested by that policy. The designation should confirm as made, including the Western Weald.'

As the Inquiry drew to a close the Inspector received a letter requesting that he should not include the Western Weald signed by a number of the major estate owners, not one of whom had previously provided evidence to the Inquiry.

We had become aware that the Inspector was rather disappointed that Bill Bryson had not made an appearance. Efforts to persuade a very busy man had not succeeded but he agreed to write a letter. This was delivered on 28 May 2008, the final day of the Inquiry and read:

'I was extremely pleased to see that the Public Inquiry had re-opened and that you have been willing to hear new evidence on why the Western Weald should be included in the South Downs National Park.

I wish to explain why I believe that the Western Weald should be included in the South Downs National Park.

When I was invited to come and see the Western Weald in Hampshire and West Sussex I must admit that I half expected to see some neat suburban landscape.

Instead I saw a gorgeous piece of England set in the heart of one of the most overcrowded parts of the planet. I saw heathlands, like Longmoor, that rivalled if not surpassed those of the New Forest. I saw fields and hedgerows unchanged since before my country was founded. All within areas not only proposed for exclusion from the National Park but also declared to be 'degraded' to have lost 'natural beauty'.

As someone who has travelled the world, who has visited all parts of this island, frankly I cannot understand where such ideas came from. Believe you me, this is a jewel; please conserve it, preserve it and confer on it the status it deserves, that of a nationally important landscape for all to enjoy as a National Park. If this beautiful landscape is left out of the South Downs National Park, I believe that it will run the risk of being picked off by developers and lost forever.

I know that the South Downs Campaign, who hosted my visit last autumn, have been working extremely hard to show you why the Western Weald should remain in the National Park boundary. I really do hope that you will now be as convinced of the case as I am!

With best wishes

Bill Bryson, President, the Campaign to Protect Rural England.'

We were enormously proud of all the evidence the various members of our team had presented. We were confident that our case was extremely strong. A bond of comradeship had built, especially over the past few months, and we had many happy memories, not least of the quality of the coffee provided by the hotel. So this day was one of extraordinary mixed emotions.

For many years we had attended our committee meetings and then gone our separate ways. By being present at the Inquiry for so many long days, we had much more opportunity to bond and to enjoy lunches together. We regularly went to an excellent pizza house close to the pier. Indeed, we were such regular customers that on the first day of the re-opened Inquiry the manager said that he had been expecting us. But on this day with so many members present, we decided to splash out and had our final lunch at a superb fish restaurant. We then departed wondering what the outcome of all our collective efforts was going to be.

For the rest of the year and into 2009 we waited nervously for the Inspector's report. Our anxiety was not restricted to the contents of the report but also how and when Defra would respond: we were all too conscious that a General Election was fast approaching.

We needed to keep the South Downs at the forefront of their minds and Chris Todd came up with the perfect answer: we would print thousands of postcards depicting the areas we wanted to see in the National Park: the chalk hills, the Western Weald, the town of Lewes and the village of Ditchling and then get people to send them into Defra. In reply to a Parliamentary Question by David Lepper in February 2009 the now Minister for National Parks, Huw Irranca-Davies stated that to date Defra had received around 16,000 of our postcards. We released a press statement announcing that we had been overwhelmed by the public response and that *'We want Defra to receive even more postcards by the time the campaign finishes. That way the Government can be left in no doubt as to the strength of public support for a South Downs National Park including Lewes, Ditchling and the Western Weald.'*

A much larger version of this postcard signed by Bill Bryson, Brian Blessed, Ben Fogle, David Dimbleby and Floella Benjamin was taken to the Defra headquarters in London's Smith Square to be presented to the Secretary of State, Hilary Benn. Bill Bryson and Michael Mates MP joined us and gave interviews to the assembled press and television reporters. The delivery of the giant postcard to Defra attracted media attention both in London and the South Downs. Margaret Paren was photographed by the local press boarding the train in Liss with a giant postcard rather awkwardly carried under her arm to the puzzlement of fellow passengers. The event also attracted the interest of Sir Peter Bottomley, the MP for Worthing West who was, at the time, vociferously opposed to the National Park. He appeared, seemingly from nowhere, and after having a confrontation with me, he proceeded to give interviews to the TV cameras. We countered by holding placards in favour of the National Park above his head.

Some years later, at a CPRE Hampshire event, Margaret Paren was approached by someone who had worked in the post room in Defra at that time. He told her that by the time the announcement was made the basement of their building was filled with trestle tables covered in postcards. Hilary Benn had personally visited the basement to see the assembled mass.

CHAPTER SIXTEEN

VICTORY AT LAST

On 31 March 2009 the Secretary of State, Hilary Benn, accompanied by his minister, Huw Irranca-Davies, came to the South Downs to make a public statement regarding the National Park.

We had been given a few days warning as CNP had been approached by Defra and told an announcement was to be made at the Weald and Downland Living Museum and that the Secretary of State had asked that some members of the South Downs Campaign should be present. We concluded that it was unlikely Hilary Benn would be travelling to the South Downs to announce that there would be no national park and it would be somewhat crass of him to announce at a museum dedicated to the Weald, as well as the Downs, that it would be a chalk-only national park. We therefore surmised that the Western Weald had been included. However, there was no certainty of this, and Defra had declined to provide anything save that all the information would be up on their website by 8am on the day. We decided that I would be at the museum and that Margaret Paren would be there too and would handle any media.

Around Midday the Ministers arrived having briefed local MPs beforehand. Huw Irranca-Davies commented that the most vociferous opposition had come from Nick Herbert, the MP for Arundel and the South Downs. We were surprised to hear that as of all the local Conservative MPs he had been the only one from West Sussex to engage with us and listen to our reasons for wanting a national park.

As we had anticipated, Hilary Benn announced that it was his intention to create a South Downs National Park that would stretch from Eastbourne to Winchester and would include the Western Weald, all its towns and villages and Lewes and Ditchling. He also proposed to consult on six additional areas including Alice Holt Forest in Hampshire and Castle Goring, just outside Worthing, neither of which we had expected to see included. However, the Order confirming the Designation Order could not be signed until a consultation on the additional areas had taken place and a proper decision made. He said:

'I have decided that the South Downs, including the so-called Western Weald, should be made a National Park. National status can be a real boost

for the economy, attracting new visitors, businesses and investments, but above all, the South Downs wonderful countryside will be protected forever for the enjoyment of everyone.'

This announcement once again created mixed emotions within the Campaign. On the one hand there was great joy coupled with a sense of wonder that the Campaign had actually done it, there was going to be a South Downs National Park with boundaries we could support. On the other hand, we felt deflated: the campaign was over: what next? After so many twists and turns on the way, we were only too aware that there was still some way to go and, with an election pending, the whole project just might be in jeopardy after all.

Nevertheless, all involved have lasting memories of the day. Here are just a few of them.

Margaret Paren remembers it vividly:

'The day was cold, grey and misty. I was accompanied to the museum by Karen Burney, the communications consultant to Hampshire CPRE. Just after 8am I phoned Chris Todd from the car to find out what we had got. He said the Defra website did not yet display the information. With ten minutes to go before my first interview, in desperation I phoned Chris Fairbrother at Natural England who, I thought, must have advanced information. Far from it. He had just seen the Inspector's report and was working through the narrative painstakingly. We deduced from the hectarage of AONB land lying outside the boundary that the Western Weald was in, but it wasn't immediately apparent what had happened to places like Woolmer Forest and the Bentley Nib which were not in the AONB. He did, however, confirm that Lewes and Ditchling had been included. Armed with this information I undertook a flurry of interviews that lasted up to the time of the arrival of the Ministers.

Based on our assumption that the Western Weald was included we had decided to celebrate by presenting a young oak tree to each of the parishes in the area that had supported the Campaign. The trees were purchased from a tree nursery near Winchester that guaranteed native oaks. They were stored in an outbuilding at Andrew Shaxson's farm in Elsted. The plan was then to plant the Elsted tree adjacent to the cricket ground, which would provide a photogenic backdrop for the assembled press photographers. To demonstrate the breadth of support for the Western Weald campaign we would display all the oak trees with a large tag on each identifying the name of the Parish concerned.

After the Ministerial announcement Karen and I were setting off from the museum to meet fellow campaigners for the tree planting ceremony. As we were leaving the museum Karen spotted some of the BBC 24-hour news crew looking a bit dejected. They explained they had been up before dawn, had had nothing to eat and didn't know where to get food. Karen suggested they follow us to Elsted where there was a renowned local pub.

At Elsted we parked at the cricket ground and pointed the TV crew in the direction of the Three Horseshoes. By the time we and the local press had assembled, the TV crew had returned from their meal and they inquired what was going on. And, so it was, that I and a group of campaigners headlined BBC1's 6 o'clock news planting a tree and drinking champagne.'

Meanwhile over in Lewes they had heard that the town had been included. Robert Cheesman recalls:

'Paul Millmore had heard from contacts in Natural England of Lewes' inclusion and he and I hastily arranged to meet in the gardens at Southover Grange later that morning where we held a private celebration. Later in the day the Mayor of Lewes organised a small drinks reception for members of the Campaign and local councillors. It was a happy occasion and we looked forward to the establishment of the National Park and its Authority.'

Not to be left out, at 1.30pm local residents in Ditchling were popping the local bubbly on Lodge Hill that sits just above Ditchling to its west providing sweeping views across the village to the Downs. After such a long battle to have the village included, residents were ecstatic at the news and a large contingent had their photos taken around the giant postcard that was doing the rounds that day. There was a twist though. Residents thought they were celebrating successfully persuading the Inspector to include Ditchling. Little did they know their efforts had been in vain as, apart from agreeing to include Lodge Hill, he wanted Ditchling out. Fortunately for them Hilary Benn, who knew the area well having studied at the University Sussex, was keen on its inclusion and insisted on visiting the area to judge for himself. However, he was told that he would have to produce a rationale for overturning the Inspector's recommendation. This was the moment when the report that had been produced for the first part of the Inquiry really proved its worth, because Defra civil servants dug it out and made great use of its contents.

Elsewhere, celebration was distinctly more muted. West Sussex County Council said that it would press Defra to pioneer a new type of national park for the South Downs. It would push for a tailor-made national park authority to make sure that major decisions, such as planning, were decided by locally elected members, accountable to local communities and not Government appointees. To his credit, Council leader Henry Smith said:

'We are prepared to accept the decision to include the Western Weald, although I must stress that this does not give the area any greater protection in law than it already has as an Area of Outstanding Natural Beauty. There are still major concerns, but what is important now is that we move on and work towards a new-style national park authority that is democratically accountable.'

Meanwhile rumblings continued including some questionable statements in an article in The Guardian. I was able to respond in a special column:

'Reading your feature about the newly declared South Downs National Park, one would think that all Britain's national parks are failing.

The simplistic comments you quote are typical of those who have opposed our campaign for national park status for the past 20 years: "The park would consist of one giant sheep field with no boundaries"; "Our members, who tend to be modern commercial farmers rather than the yokels and milkmaids of rural fantasy were horrified by this; This very thriving part of the south-East (would be) reduced to tea shops and car parks".

The reported "deep rumblings of discontent" from those opposing the park mostly come from a small political elite who have been unwilling to listen to their electorates and, as one local paper eloquently put it, whose efforts have smacked of being more concerned in protecting their own empires than protecting the South Downs.

The National Farmers Union suggests that "other national parks have large expanses of wildernesses whereas the South Downs is highly managed". In fact, nearly all the land in Britain's national parks is farmed, from the sheep and pony grazing on the uplands to the stone-walled hay meadows of the Yorkshire Dales. Likewise, in the South Downs, nature conservationists and the body that looks after the Areas of Outstanding Beauty are restoring the neglected heathlands and ancient chalk grasslands by reintroducing grazing regimes.

Then there are the fears about the hordes of day-trippers and coach parks. Yet most of the 40m visits a year to the South Downs are from local people. The greatest threat from visitor numbers is likely to come from the increasing population in the south-east. All the more reason why there should be a national park authority (NPA) with the resources and expertise to manage access, and a statutory duty to give priority to conservation over recreation.

The loudest opposition cries come from those, including MPs, who claim that the South Downs will be managed by an unelected quango or "in the hands of Whitehall apparatchiks". They conveniently ignore the 1995 Environment Act that created independent NPAs – giving county, district and parish councillors *the majority of seats. The minority nominated by the Secretary of State are appointed for their specialist and local knowledge.*

Even those councils that have been vociferously opposed to the park are now expressing their willingness to co-operate in creating the best possible arrangement for managing it. We look forward to our new NPA with the ability to plan for the long-term, with the funding and the expertise to deal with the pressures that will inevitably face one of the most iconic and much-loved areas in Britain.'

The announcement of the new South Downs National Park was accompanied by the publication of the Inspector's report, which had been delivered to Defra on 28 November 2008. We were delighted by what we read and glad to see that all the hard work had been worthwhile. Of particular note was the Inspector's recognition that the Meyrick case and the NERC Act amendments had allowed for him to take a different approach from that taken by the Inspector in the New Forest. Thus, a fresh look at the case for a wider national park was possible. He also said that the mass of new evidence had now satisfied him of the quality of the Lower Rother Valley. So far as the 'A3 corridor' and Upper Rother Valley was concerned he accepted that in the final analysis it should be protected and that if the Secretary of State decided differently a new boundary setting exercise was necessary.

What had changed between the two parts of the Inquiry to enable such a dramatically different outcome regarding the Western Weald? We believed there were four reasons.

First, at that time, the Planning Inspectorate was understandably keen to ensure consistency between the decisions taken by their Inspectors. Once the New Forest National Park Inquiry's Inspector had chosen to exclude land 'not of New Forest character', and that decision had been accepted by Defra, the South Downs Inspector's hands were, in effect tied, unless evidence was presented to him that clearly led to a different conclusion. The Inspector acknowledged this point in his final report when he said that the changes to the NERC Act allowed him to reach a different conclusion to that in the New Forest.

Secondly, the structure of the first part of the Inquiry was based on Local Plan inquiries which allowed only those against what was proposed to have a voice. The opposing councils chose to make their case for a smaller national park primarily on the basis of landscape character, that only the chalk landscapes could qualify. Thus, the case for the inclusion of the Western Weald was not heard, no evidence on its quality or linkages to the chalk. And, presumably, because Defra had accepted the findings of the New Forest Inspector, no evidence was heard to challenge the legality under the 1949 Act of the landscape character argument. Not only was the Campaign excluded from making a case for the Western Weald but so too was Hampshire County Council. The plain fact is that the overwhelming evidence presented at the second part of the Inquiry was the first occasion the Inspector had heard the case for inclusion of the area.

Thirdly, the opposing councils not only agreed to hearings on the Western Weald in the second part of the Inquiry but they then produced no new evidence themselves, a fact the Inspector alluded to during one of the hearings.

Finally, the second part of the Inquiry allowed the Landscape Assessor's reports to be thoroughly examined and their weaknesses exposed. Thus, the challenge to the quality of the Western Weald used to back up the Inspector's findings from the first part of the Inquiry fell.

It is sobering to think that had there not been a judicial review of the New Forest National Park Confirmation Order and a subsequent amendment in the NERC Act, the Secretary of State's decision would have been based on the first part of the Inquiry alone and the South Downs National Park would now be a smaller and poorer place. It is even more sobering to think that had the Landscape Assessor's report not been scrutinised in public the whole of the Western Weald, or large swathes of it, could have lost protected status, the largest de-designation ever of a nationally protected landscape.

There was still work to be done. We had to respond to a consultation on six areas the Secretary of State was minded to include but had not previously been the subject of consultation. The boundary group reformed and to start with reviewed the areas to see if we had any comments to make on either the supporting evidence or the proposed boundary. They didn't need to visit three of the areas as they were proposed in order to create a better boundary and were fairly straightforward.

They started their day of site visits at Alice Holt Forest, which was the largest area in question. The Countryside Agency had included the western part within its Area of Search. The Campaign had examined the forest as a whole a long time ago but it had been decided that it was not a priority. However, one of our number, John Templeton, had always been an advocate of Alice Holt Forest being included, and had the quiet satisfaction of knowing he had made a personal submission to the Inquiry making a case for it to be a gateway to the National Park. The group reported back that though they thought the Campaign should support the inclusion, they recommended a modified boundary to exclude

'Birdworld' and the adjacent garden centre.

After lunch at the Three Horseshoes in Elsted they proceeded to Castle Goring, which presented a greater challenge. The house itself, a Grade 1 listed building, is decidedly quirky and thus interesting, and the extensive grounds have a clear visual connectivity to the chalk hills despite being separated from them by the A27. Part of these grounds are protected as ancient woodland, along with further woodland outside the perimeter to the east of Titmore Lane. The Countryside Agency had included the land to the east of Titmore Lane in the designated boundary but excluded Castle Goring. The Inspector had heard the case for inclusion of the whole area at the first part of the Inquiry from the Worthing Society, others had called for its exclusion. Despite criticising the Countryside Agency's evidence, he accepted its reasoning for the most part and consequently recommended that the whole area should be excluded largely on the grounds that there was no public access. In his second report he had reversed his recommendation and proposed the inclusion of the whole area 'on balance' despite the doubts about public access. Our group had continued doubts about access, particularly given the Appeal judgment on the Meyrick Case. In the end we decided to stand back and let the Worthing Society make the case.

Finally, the group visited the fields near Plumpton. By this stage, they had full ownership of the Countryside Agency's boundary setting criteria and were rather appalled that one of the fields with discernible detractors like a power line and the detritus of 'horsey culture' should be thought to merit inclusion. We therefore decided to oppose this at the re-opened Inquiry.

The Campaign also saw this further consultation as an opportunity to take up errors in the decision letter and Inspector's report. In April we wrote to Defra outlining a number of concerns, not least the fact that the boundary at Offham had split the village, the boundary at Southwick Tunnel had been incorrectly altered and that the Inspector had made an error in his report stating that the objection to the Gote Farm extension had been withdrawn. In reality only 3 paragraphs of this objection had been withdrawn. Left uncorrected, this could have opened up the decision to a legal challenge.

At the same time the Campaign worked closely with the group 'Keep The Ridge Green' to rectify what appeared to be a boundary error at Green Ridge on the outskirts of Brighton. It appeared that the Inspector had been confused and viewed the area as being an extension of Toads Hole Valley (a neighbouring area but having no visual connectivity with Green Ridge) which he had excluded. This was a harder nut to crack and required a strong and vociferous campaign

Bill Bryson and Margaret Paren outside DEFRA headquarters in London
©*Chis Todd*

Hilary Benn displaying the Confirmation Order with Margaret Paren, Robin Crane, John Templeton, Owen Plunkett and others
©*Chris Todd*

backed by the local MPs David Lepper and Des Turner and Brighton & Hove City Council to raise its profile politically. This was important, not least because Green Ridge risked not being reconsidered as it was not one of the six additional areas. The Campaign concentrated on producing a technical basis on which to challenge the decision including highlighting new evidence. It also argued that the site, being adjacent to the A27 embankments, could be considered as part of the consultation on the six additional areas of which the A27 embankments was one.

The Inquiry re-opened for one day in August in the Chatsworth Hotel to consider the outstanding issues. It was a relaxed, almost jolly event in contrast with the many grim months we had experienced there before. The Inspector entered into the mood of the proceedings by commenting that this was the first time he could recall the South Downs Campaign actually opposing land being included in the National Park. He accepted our arguments for a boundary modification at Alice Holt Forest but included the field near Plumpton. He included Green Ridge.

In September, we submitted our response to the consultation on the proposed National Park Authority membership. In it we expressed our concern at the geographical deficit with regards to Hampshire created by the suggested membership models. We therefore proposed a thirty-three seat authority with extra seats being given to the Hampshire authorities. This model was rejected by Defra and all local authorities were treated equally, with one member each, creating a National Park Authority with a maximum membership of 28.

Our remaining disquiet at this point was about timing. How soon would the Confirmation Order be signed? We were concerned at what seemed very slow progress in putting together an establishment team to set up the National Park and had no idea of the timing for the necessary secondary legislation to set up the National Park Authority. Would the National Park come into existence before the General Election or would all our hopes be dashed?

CHAPTER SEVENTEEN

THE FINAL STAGES

On 12th November 2009 the Secretary of State, Hilary Benn came to Ditchling to sign the South Downs National Park (Designation) Confirmation Order.

We were given the date for the signing of the Order but there was some confusion as to where this would be. Originally it was to be Ditchling Beacon, chosen by Hilary Benn because it had been one of his favourite places when at the University of Sussex. However, the weather forecast for the day was truly grim with heavy rain and high winds forecast for the area. After some toing and froing with Defra we were given the name of a tea-room in Ditchling which, we were told, would house the numbers invited to the signing ceremony. When we got there, however, we found the tea-room was a series of smaller rooms none of which were large enough to house everyone. We were in the process of eating a breakfast of bacon rolls when Hilary Benn, a vegetarian, arrived.

In no time at all the signing ceremony was switched into the walled courtyard behind the tea-room since the weather, although windy and changeable, was not as bad as had been forecast. What with the invitees, the press and TV cameras it was a very tight fit. The leader of the newly formed establishment team, later the interim Chief Executive of the South Downs National Park Authority, Richard Shaw, had wisely arranged for some local farmers and land managers to join the celebration and for a supply of local sparkling wine kindly donated by Ridgeway Wines, which was located just up the road. Thus, the teetotal Hilary Benn was photographed marking the occasion with a glass of bubbly in his hand.

The signing of the Confirmation Order was not just a ceremonial affair. Six copies had to be signed as well as the Orders abolishing the Sussex Downs and East Hampshire AONBs. It was legally necessary that all these documents were signed on the same day. The task of getting the job done fell to Chris Fairbrother of Natural England: it was fitting he should be in attendance on such a historic day.

The Defra officials had worked hard too to get everything ready. They brought with them many copies of the Confirmation Order maps, which were enthusiastically grabbed by those present.

Hilary Benn signed one copy of the Confirmation Order in the courtyard before hastily retiring into the tea-rooms. I can't recall who it was who first asked him to sign one of the maps, I think Chris Todd, but many of us made a queue to get our copies signed as well. Then the owner of the tea rooms asked him to sign a tablecloth, then a wall and then a tea cloth for her mother. All of this signing was done with great good humour. We added to the chaos as Chris Todd had brought along one of our giant postcards and was encouraging everyone present to sign the back of it as a memory for all those who had been involved with the Campaign. No one wanted to be left out of the celebrations and it was indeed a festive occasion. In the midst of it all sat Chris Fairbrother with his head in his hands. 'He's signing everything except my Confirmation Orders' he said. Indeed, Chris had to get on his motor bike and drive up to the Defra offices later in the day to get all the signatures he needed on the documents.

The Order confirmed that, after objections and representations were duly made, and a local Inquiry held, the South Downs National Park (Designation) Order made by the Countryside Agency in 2002 would take effect on 31 March 2010. On that day the two AONBs would be de-designated.

The scale of our success was overwhelming. Nearly all the additional areas we had proposed were in the National Park including the substantial area of the 'Bentley Nib'. Hilary Benn had agreed with all the Inspector's recommendations on the boundary as well as including the six additional areas that he had proposed himself. The whole of Green Ridge and Offham village made it into the boundary as well.

We had been extremely lucky to have, as the Secretary of State, Hilary Benn, whose personal commitment undoubtedly ensured the latter stages of the process was pressed forward as speedily as possible. His personal recollections of his engagement are worthy of repetition here.

'I walked into my office at Defra one day to find an unusually large pile of documents on the desk. Upon enquiring what they were, I was told that it was the second report of the Inspector on the South Downs National Park. I knew that there had been long disputes and debates about the idea of the Park but I had no idea that the report was about to arrive.

My officials took me through the decisions I would have to make not only about the Park as a whole but also about a number of specific areas of land where the Inspector had left it up to me to decide whether or not they should be included in the National Park.

I had a look at the maps and said to my officials – who were wonderful throughout the whole process – "Well, I had better go have a look". They appeared somewhat taken aback and asked me what I meant. I explained that if I was going to take a decision on these individual pieces of land then I wanted to go and see them. "But Secretary of State" they said: "You can't do that." "Why not?" I replied. "Because what happens if you talk to someone?"

They then pointed out that because I would be acting in a quasi-judicial capacity, I had to avoid any suggestion that someone had been able to influence my decision while I was considering the Inspector's report. So we reached a very British compromise. It was agreed that if someone said good morning to me then I could say a cheery good morning back and that the same would apply in the afternoon.

I will always remember the day that we travelled in a minibus from one end of the proposed Park to the other to look at the various locations and landscapes. It was a wonderful opportunity to see the many delights and splendours of the Western Weald and the chalk Downs, or as the Park Authority puts it so eloquently "the rolling hills, glorious heathland, river valleys, ancient woodland, thriving villages and market towns, and the iconic white cliffs.."

I decided to include almost all of the sites, including Ditchling for which I have a particular fondness. I had visited it before and seen the way in which Ditchling Beacon is framed between the buildings at the crossroads. My officials told me that if I was going to include Ditchling then I would have to be satisfied that the land to the north of it was of 'national park quality'. So we drove north for about a mile whereupon I declared myself to be so satisfied. We turned round and as we approached the crossroads I declared that Ditchling would indeed be part of the National Park. I also had the idea that when it came to signing the formal legal documents, it would be to there that I would return.

And so it was, on a slightly damp November morning just over 10 years ago in 2009, that we gathered in the back garden of the tearooms in West Street for the signing of the official papers establishing the National Park, as Ditchling Beacon peered approvingly at us over the back wall.

With the flourish of a pen, the hopes and dreams of all those people who for years had campaigned to create a National Park in the South Downs were realised. We owe an enormous debt of gratitude to each and every one of them for without their passion, effort and sheer determination, it would not have happened. All of you should look back on your magnificent achievement with great pride.

Creating the South Downs National Park is one of the things I am most proud of from my time as a Minister. It is not often that you are given the chance to preserve beauty for succeeding generations. I was also very conscious that I was walking in the footsteps of my predecessors at Defra who had worked so hard to bring this about - including Michael Meacher and Margaret Beckett. So the night before the tearoom signing, I sent them both a map of the boundaries of the new Park to thank them for everything they had done.

Thus was the South Downs National Park born. It was the last on the original list of beautiful and precious places drawn up by Sir Arthur Hobhouse in the late 1940s which led to the passing of the National Parks and Access to the Countryside Act of 1949.

At the end of his speech moving the Second Reading of the Bill in the House of Commons, the Minister Lewis Silkin said this:

"It is perhaps a reason for our country's greatness that in a difficult period like the present we are not afraid to set aside time and energy for the practical measures needed to help people to enjoy these beautiful areas....... Now at last we shall be able to see that the mountains of Snowdonia, the Lakes, and the waters of the Broads, the moors and dales of the Peak, the South Downs and the tors of the West Country belong to the people as a right and not as a concession. This is not just a Bill. It is a people's charter—a people's charter for the open air, for the hikers and the ramblers, for everyone who loves to get out into the open air and enjoy the countryside. Without it they are fettered, deprived of their powers of access and facilities needed to make holidays enjoyable. With it the countryside is theirs to preserve, to cherish, to enjoy and to make their own."

He knew - we all know - that the South Downs is a special, indeed, a magical landscape.

Long may it remain so!'

Celebrations were arranged for later in the day. In Lewes the mayor was host and in Midhurst I was joined by the Town Council chairman Colin Hughes, who had worked so hard for our cause, Andrew Shaxson (Chichester District Council and Chair of Elsted Parish Council), John Andre (Woolbeding) and Jacquetta Fewster, Director of the South Downs Society.

That evening I drove up onto the Downs at Treyford to enjoy the rolling chalk hills and the spectacular view across the Western Weald. I found it hard to take in the reality that our South Downs National Park was there in front of me and that it would cover some 632 square miles (1637 square kilometres), making it the third largest national park in England.

We had further good news that day. Paul Millmore, in particular, had been a great proponent of having an open boundary to the sea so that, he hoped, the National Park Authority would have some leverage over what would happen to the rich marine environment offshore of the Heritage Coast. There was no certainty of this: a number of existing national parks had open borders to the sea but had no consequent locus beyond the land. Nevertheless, the case had been made with great vigour at the Inquiry by Paul and Phil Belden and the Inspector had been persuaded. Subsequently Defra had not accepted his recommendation. Paul was incensed and threatened to take legal action against Defra paid for out of his own pocket if necessary. Some of our Executive were alarmed by this: having a land-based national park was what they had strived for, they didn't want the process derailed at this late stage. Subsequently Paul Millmore, Phil Belden and Chris Todd went to see Defra officials who provided reassurance that the marine environment would be protected under separate legislation. And so it proved to be: the Marine and Coastal Act of 2009 declared 'Beachy Head East' and 'Offshore Brighton' as Marine Conservation Zones (MCZ) These areas were designated with the aim to protect nationally important, rare or threatened habitats and species. The Bill received Royal Assent on the same day the South Downs National Park Confirmation Order was signed.

Meanwhile Andrew Tyrie, MP for Chichester, did his best to spoil the party with a swingeing statement. He described the decision as a *'frontal attack on local democracy'* which took planning decisions out of the control of locally elected councillors. In response Hilary Benn said, *'I do recognise that the National Park is different in character because it has the biggest population. I do want the National Park Authority, where possible and appropriate to enable planning decisions to continue to be taken by the local authorities. National Park status can be a real boost for the local economy, attracting new visitors, businesses and investment. But above all, the South Downs's wonderful countryside will be protected for ever for the enjoyment of everyone.'*

On 10 December 2009 the Secretary of State announced that the new National Park Authority would consist of 27 members and that the local authorities would have a majority with 14 members (with Worthing and Adur Councils sharing a seat). There would be 7 members appointed by the Secretary of State to represent 'the national interest' (normally local people) and 6 parish council members (2 from each county). The Secretary of State expressly requested that the Authority should delegate as much development management as possible.

On Tuesday 2 February 2010 the South Downs Campaign held its final AGM at the Minerva Theatre Chichester at which its Executive's proposal to disband the organisation was agreed. A number of organisations though wished to remain in touch on South Downs matters and it had been decided that a loose federation should be set up to cater for their needs. Thus it was resolved to set up a South Downs Network.

There then followed a celebration chaired by the Mayor of Chichester, who welcomed us to his city on behalf of the city council, which strongly supported the South Downs National Park. He said it was the first time as Mayor that he had been asked to close something rather than to open it and declared that the South Downs Campaign was well and truly closed!

The Chairman's final address (Robin Crane)

'Thank you all very much for coming today. I hope that you have enjoyed having the opportunity to reflect on our history, to look forward to the future and to celebrate the establishment of our South Downs National Park.

We have so much to celebrate; beyond doubt the South Downs Campaign has been an extraordinary and unprecedented phenomenon. Never before have so many national, regional and local organisations and numerous individuals worked so closely together.

Several times I have been asked how it is that we have managed to survive for such a long time without falling out. The glib answer is to say that it was because we were all striving together with just one single aim. However the establishment of a national park is a complex subject. There have been plenty of opportunities for disagreements to develop during our long journey. My answer is that we have succeeded because we have been extraordinarily fortunate in having such a splendid team of individuals working together and in Chris Todd we have had a consummate diplomat who has been our main contact with our supporters.

I believe that we should all be proud of the fact that throughout our campaign we have conducted ourselves with great integrity and honesty. We have always let the facts speak for themselves. We have never dealt in misinformation or in any way attempted to mislead the public.

When I set out on this enterprise, I never expected that politicians would loom so large in our affairs. I won't spoil the party by repeating my well known opinions of many of our opponents. But I must mention those who have played a supportive role: first and foremost Michael Meacher who overcame difficulties to set the designation process in motion. I am really sorry he could not be with us today. At the local level the MP's from Brighton, Dave Lepper and Des Turner have taken a very close personal interest in our campaign and have frequently appeared at our conferences. Celia Barlow has also been a keen supporter. Michael Mates, James Arbuthnot, Chris Huhne and, latterly, Norman Baker have given us cross-party support. In the House of Lords, where most of the parliamentary debates on the South Downs have taken place, Lord Bassam and Lord Judd have frequently acted on our behalf.

In the political context we must also thank one of our longest serving and staunchest supporters John Carden, whose family have had a long association with the conservation of the Downs. John has beavered away quietly and very effectively behind the scenes of the Labour Party.

We have lost count of the number of Ministers we have had to deal with during our campaign! However, to have Hilary Benn and Huw Irranca Davies finally signing off the Park has been the icing on the cake. We could not have had two more warm and enthusiastic Ministers to lift us off the launch pad. I know that both of them would have loved to have been with us here today.

A lustre to our Campaign has also been added by the high profile support of Bill Bryson, David Dimbleby, Ben Fogle and Honeysuckle Weeks. Bill Bryson gave us much valuable time here and in London. Brian Blessed has also been hugely supportive but totally uncontrollable. Libby Purves kindly visited us when CNP arranged for a press launch with her and Brian Blessed on the precipitous Mount Caburn.

Nothing could have been achieved without money. We must thank all our organisations and many individuals for so generously supporting us. We are also grateful to the Esme Fairbairn Foundation, the Kleinwort Trust and the Gatliff Trust for their charitable donations over several years. Without them we would not have survived.

Over the course of a twenty- year campaign there have inevitably been people close to us who have sadly passed away. Kath Worvell was a staunch supporter for many years, as were Sheila Schaffer and Denis Payne. David Clegg took the lead in fighting for the inclusion of Rowlands Castle. Dr Francis Rose was a strong advocate for the national park and provided evidence for the public Inquiry. More recently we lost Nigel Paren. He not only gave Margaret tremendous support but was very active himself in fighting for the inclusion of Liss, Longmoor and Woolmer Forest. Now we have sadly lost John Venning. He was not a member of our executive for all that long, but with his charisma, humour and wise advice he made a great impact upon all of us. We do hope that it will be a small consolation for all these families in knowing that the South Downs National Park will serve as a memorial for them for ever.

I must now pay tribute to a few individuals who have played a prominent role in our campaign. Firstly, Phil Belden, Paul Millmore and Amanda Nobbs who were founder members of the campaign. Amanda was Director of the Council for National Parks and our pivotal player in the early years. Phil continues to provide excellent support. Paul has been our most passionate supporter whose enthusiasm on our executive has never been daunted. In the early years. Andrew Lyall was our extremely efficient honorary secretary.

Finally I must give very special thanks to the three people who have working most closely with me over many years on an almost daily basis.

Margaret Paren has been a huge asset in our campaign since 2002. It was clear from the beginning that she was an extremely able and knowledgeable person who was dedicated to the South Downs. With her very sharp, analytical mind Margaret has kept us all on our toes ever since. Initially Margaret led our Boundary Working Group. She was also very closely involved in much of the preparation and editing of the 80 written representations and proofs of evidence that we submitted in the first part of the public Inquiry. Her greatest contribution was when we were all under extreme pressure when given only a few weeks to prepare our case for the re-opened Inquiry. Margaret put in a huge amount of work into our defence of the Western Weald. With inspired leadership she brought together a great team of supporters who also performed wonders in the little time that was available to assemble the evidence that demolished the opposition.

Margaret, we thank you for everything and sincerely hope that you will be successful in your bid to become one of the Secretary of State's nominees on the National Park Authority.

Ruth Chambers from the Campaign for National Parks has been hugely influential at the national level and her knowledge of Britain's other National Parks has been invaluable. I am delighted that she is already acting as the first chairman of our successor body The South Downs Network. We urge all of you to give the Network your wholehearted support.

And now to Chris Todd. His contribution to our campaign has been truly outstanding. He was already a successful campaigner for the National Park before he became our Campaign Officer nine years ago. Since then, he has carried out a multitude of tasks in administering and promoting our campaign, all done with great efficiency and to the highest standards. His work rate has been phenomenal. His input into the public Inquiry was highly professional and ensured that all our evidence was immaculately presented. Chris, our campaign couldn't possibly have been so effective without your dedication, and we are all truly grateful. We give you our very best wishes for the future.

And so we come to the close. What we have achieved is monumental. We have secured a firm base but we have not reached Utopia yet. The future of our South Downs National Park will largely rest on political decisions, such as in agriculture, forestry, housing development and nature conservation. We have enjoyed overwhelming public support and we must take advantage of it. It is now up to you all to keep your feet on the pedals and act robustly to make certain that the resources required are made available to support local communities and to conserve and enhance our wonderful inheritance for future generations.

My very best wishes to you all.'

It had been a hard fight against determined opposition. In my address I concentrated on the contribution of the members of the Campaign and the politicians who had played a key part in the process. But it is also the case that we were very much supported by the media, particularly the local press and radio stations and the regional news programmes. They kept the Campaign in the public eye over many years. Above all, we had the support of huge numbers of the public. To my mind, that is the key: no national park can be established and prosper if its communities are at odds with the designation.

On 31 March 2010 the South Downs National Park was established. A day later, the shadow National Park Authority came into being. Margaret Paren became a member of the Authority and its founding Chair. She received an OBE for services to the Environment.

Barely six weeks after the establishment of the National Park the General Election took place and the coalition government was formed, bringing in a programme of austerity. It had been a very close call.

THE SOUTH DOWNS CAMPAIGN

A TRIBUTE BY LEN CLARK CBE (1916-2019)

'The Campaign generated a remarkable amount of enthusiasm and uncovered much expertise. But perhaps the most remarkable feature of the Campaign was the harmony and good humour maintained throughout. As I have commented often, all too frequently groups with a common goal fall foul to friction en-route: "enthusiasts beget zealots, zealots beget bigots, and bigots beget bloodshed". Not so with the Campaign. There were obviously occasions for differing views on tactics and methods, but these always found acceptable outcomes.

The Campaign would have been ineffective without the dedication and hard work of the whole team, and its 160-odd affiliates. Patience, hard work and determination have been the hallmarks, but so have tolerance spiced with good humour. At times most of us have empathised with both Sisyphus striving to lodge the stone at the top of the hill and with Moses, trying to take the tribes of Israel to the Promised Land.

The National Park is not a panacea for all planning and land use problems, but it is the best, and much deserved tool for the job.

To have played even a minor a part in this journey has been a great privilege.

Despite the inherent strength of the Campaign it is difficult to imagine what would have happened to it without its Chair, Robin Crane, who came in from the 'wildlife stable' and was chair throughout. It might well be said that he did not take it over but the Campaign took him over. As he also acted as Honorary Treasurer, our debt to him was compounded. Seeking financial backing, answering letters in the Press, handling media and writing papers – these were only some of his daily diet. But at the endless meetings he showed a patience with setbacks which I think would have surprised himself in retrospect. It should be added that the patience was always coupled with determination to press on. But there was a dividend for him, in that he knew he had a loyal team. The Honorary Doctorate awarded to him by the University of Sussex was richly deserved.'

ACKNOWLEDGEMENTS

First and foremost, I must thank Margaret Paren, who has spent a huge amount of time going through this whole story, meticulously checking the facts and editing the text as necessary. The enormous amount of work involved in presenting our case at the two parts of the Public Inquiry was led by Margaret and she has been particularly helpful in putting together this part of our history.

I must also thank those who gave me support and encouragement throughout the twenty years of the South Downs Campaign. Top of the list must be my late wife Wendy Crane, who never complained about my work in our retirement years although, with tongue in cheek, she did often describe herself to our friends as being a 'South Downs Widow'.

Margaret and I give special thanks to Sarah Hartnell, who gave much time to casting her independent and legal eye over the final draft and made many helpful comments and suggestions.

Special thanks are also due to Mike and Pippa Bass, Phil Belden, the Rt. Hon. Hilary Benn MP, Robert Cheesman, Brian Cheater, the late Len Clark, Vicki Elcoate, David Lepper, Pat Leonard, Emma Marrington, Christopher Napier, Amanda Nobbs, Minette Palmer, Owen Plunkett, Emily Richmond, Marion Shoard, Alison Tingley, Chris Todd and Tony Whitbread for their encouragement and helpful comments and/or contributions to the text. I am also most grateful to John Templeton who provided much valuable information which he gleaned from the detailed diaries which he kept throughout his long involvement with the Campaign. Thanks also to the Hobhouse Archive and to David Wilkinson, author of *Fight for it Now*, the story of John Dower's struggle for national parks, for providing some unpublished material on the National Parks Committee's visits to the South Downs in 1946.

In weaving together the remarkable history of the South Downs Campaign I spent many, many weeks going through every minute and document. It is inevitable that, in the pursuit of the truth, I have sometimes lifted words straight into my narrative that were not written by me. Where possible I have identified the author or organisation who originated them. Having said that, one of the greatest strengths of our organisation was our teamwork and very few documents or letters were written without the original author passing them round to colleagues for comments and corrections before they were published.

MEMBERSHIP OF THE SOUTH DOWNS CAMPAIGN (31 MARCH 2009)

National
1. Council for National Parks
2. Campaign to Protect Rural England (CPRE)
3. Friends of the Earth (FOE)
4. John Muir Trust
5. Open Spaces Society
6. Ramblers' Association
7. Royal Society for the Protection of Birds (RSPB)
8. The Wildlife Trusts
9. WWF-UK
10. Youth Hostels Association (YHA)

Regional / County
1. Badger Trust - Sussex
2. Butterfly Conservation - Sussex Branch
3. Council for British Archaeology - South East
4. CPRE Hampshire
5. CPRE Surrey
6. CPRE Sussex
7. CTC South East
8. FOE Hampshire & Isle of Wight Network
9. FOE South East
10. Hampshire & Isle of Wight Wildlife Trust
11. Hampshire Conservation Volunteers
12. LDWA Wessex
13. Ramblers' Association Sussex
14. Ramblers' Association Hampshire
15. South Downs Society
16. Surrey Under 40s Ramblers Group (SURG)
17. Sussex Amphibian & Reptile Group
18. Sussex Archaeology Society
19. Sussex Wildlife Trust

20. Sustrans South East
21. YHA South Region

Local

Parish and Town Councils

1. Alfriston Parish Council
2. Bepton Parish Council
3. Cocking Parish Council
4. Ditchling Parish Council
5. Fernhurst Parish Council
6. Funtington Parish Council
7. Greatham Parish Council
8. Hassocks Parish Council
9. Horndean Parish Council
10. Kingsley Parish Council
11. Lavant Parish Council
12. Lewes Town Council
13. Lindford Parish Council
14. Liss Parish Council
15. Midhurst Town Council
16. Milland Parish Council
17. Petersfield Town Council
18. Rogate Parish Council
19. Rowlands Castle Parish Council
20. Seaford Town Council
21. Selborne Parish Council
22. South Wonston Parish Council
23. Steep Parish Council
24. Stroud Parish Council
25. Trotton with Chithurst Parish Council
26. Washington Parish Council
27. Woolbeding with Redford Parish Council
28. Worldham Parish Council

Local organisations

1. A27 Action Group
2. Alfriston and Cuckmere Valley Partnership
3. Alfriston & District Amenity Society

4. Ashdown Rambling Club
5. Benfield Wildlife & Conservation Group
6. Bexhill Ramblers Club
7. Bishop's Waltham Society
8. Blackwater Valley FOE
9. Bricycles
10. Brighton & Hove CTC
11. Brighton & Hove FOE
12. Brighton & Hove Pensioners' Forum
13. Brighton Urban Wildlife Group
14. Buckmore Avenue Residents' Association
15. Campaign for Better Transport – East Sussex
16. Castlegate Residents' Association
17. Catherington Village Residents' Association
18. Chichester Society
19. Cycle Lewes
20. Ditchling Society
21. East Blatchington Pond Conservation Society
22. East Hampshire CPRE
23. East Preston & Kingston Preservation Society
24. Eastbourne Ratepayers' Association
25. Emsworth Residents' Association
26. Federation of Arun District Amenity Groups
27. Felpham Village Conservation Society
28. Ferring Conservation Group
29. Findon Valley Residents' Association
30. Friends of Bevendean Down
31. Friends of Hollingbury & Burstead Woods
32. Friends of Lewes
33. Friends of Stanmer Park
34. Friends of Telscombe Tye
35. Friends of Waterhall
36. Fyning, Terwick & Borden Wood Residents' Association
37. Godalming and Haslemere Ramblers' Association
38. Gosport & Fareham FOE
39. High Weald Walkers
40. Isle of Wight FOE

41. Keep The Ridge Green
42. Lewes District FOE
43. Lewes Footpaths Group
44. Lewes Trees Group
45. Liss Archaeological Society
46. Liss Area Historical Society
47. Liss Forest Residents' Association
48. Liss Village Design Group
49. Littleworth Residents' Association
50. Manhood Peninsula FOE
51. Ovingdean Residents and Preservation Society
52. Pells Amenity Group
53. Petersfield Society
54. Phoenix Action
55. Preston & Old Patcham Society
56. Railway Land Wildlife Trust
57. Arun Adur Ramblers
58. Beachy Head Ramblers
59. Crawley & North Sussex Ramblers
60. Heathfield & District Ramblers
61. Horsham and Billingshurst Ramblers
62. Meon Ramblers
63. Mid-Sussex Ramblers
64. North Hampshire Downs Ramblers
65. Portsmouth Ramblers
66. Rother Ramblers
67. Southampton Ramblers
68. South West Sussex Ramblers
69. Waltham Ramblers
70. Winchester Ramblers
71. Roedean Residents' Association
72. Rotherlands Conservation Group
73. Rudgwick Preservation Society
74. Saltdean Residents' Association
75. Saltdean Swimmers
76. Sheet Village Association
77. Slindon Common Residents' Association

78. South East Hants Ramblers' Association
79. South of High Street Angmering Residents' Association
80. Summersdale Residents' Association
81. The Angmering Society
82. The Brighton Society
83. The Kingscliffe Society
84. The Lynchmere Society
85. The Midhurst Society
86. The Round Hill Society
87. The Worthing Society
88. West Liss Residents' Association
89. Woodbury Avenue Residents' Association
90. Woolmer Forest Heritage Society

Local business associates
1. Blackdown Riding Club
2. Cycling Support Services
3. Emsworth Business Association
4. Folkington Estate
5. JCJ Pottery
6. Keith Miller Insurance Services Ltd
7. Lime Moose (Home Cooking)
8. Midhurst Tourism Partnership
9. myPetersfield
10. Treatments on the Hill (now Brighton Holistics)

Total membership = 159

BIBLIOGRAPHY

Brandon Peter	*The South Downs*
	Phillimore 1998
Clark Len	*Out of the Wind Volume 2*
	Alastair Clark 2018
Council for National Parks	*South Downs National Park – Opportunities for Enhancement 2001*
Countryside Commission	*Protecting our finest countryside: Advice to Government 1998*
Countryside Agency	*A South Downs National Park public consultation report 2001*
	Countryside Agency Publications 2001
Countryside Agency	*A South Downs National Park local authority consultation May 2002*
	Countryside Publications 2002
Crane Robin	*The South Downs National Park*
	British Wildlife Vol.31 2011
Crane Robin	*The Natural History of the South Downs National Park*
	Sussex Archaeological Society 2013
Edwards Ron	*Fit for the Future*
	Countryside Commission 1991
Hansard	Parliamentary Debates
Hobhouse Sir Arthur	*personal papers* by kind permission of the Hobhouse Archive
Hudson W H (1923)	*Nature in Downland*
	Macdonald Futura Publishers 1923
Manley John	*The Archaeology of the South Downs National Park*
	Sussex Archaeological Society 2012
Parry R N P	*The South Downs National Park Inspector's Report March 2006*
	Planning Inspectorate
Parry R N P	*The South Downs National Park Inspector's Final Report Nov 2008*
	Planning Inspectorate

Smart G & Brandon P	*The future of the South Downs 2007*
	Packard Publishing
Shoard Marion	*The Theft of the Countryside*
	Maurice Temple Smith 1980
Sussex Wildlife Trust	*A Vision for the South Downs 1993*
Woolmore Ray	*The East Hampshire AONB*
	Ray Designation Histories Series 2000
Wilkinson David	*Fight for it now*
	Signal Books Oxford 2019

INDEX

Page numbers in *italic* font indicate pages with illustrations.

A3 corridor 101, 102–3, 131, 135, 150, 152, 155
Abercrombie, Sir Patrick 29
Addison, Christopher, 1st Viscount 29
Addison, William, 4th Viscount 65, 80, 81
Addison Committee 29–30
agriculture 26, 32, 36–8, 45–6, 76, 151
  see also ploughing
Agriculture Act (1986) 37–8
agri-environmental schemes 125–6
  see also Environmentally Sensitive Areas (ESAs); subsidies
Alice Holt Forest 165–6, 168
Andre, John 172
Ankers, Steve 118, 140
Arbuthnot, James 135, 175
Areas of Outstanding Natural Beauty (AONB) 32–3, 35, 36, 46–7, 66–7, 70–3
  see also East Hampshire AONB; Sussex Downs AONB; Sussex Downs Conservation Board
Arun, river 23, 101
Arundel 121, 131, 140
Ashbrook, Kate 80, 86, 89, 109
Ashdown, Paddy, Lord 61

Baines, Chris 67
Baker, Norman 63, 140, 175
Bangs, Dave 62, 76, 77–8, 113
Barlow, Celia 175
Bassam, Steve, Lord 57, 175
Beaumont, Tim, Lord 80
Beckett, Margaret 172
Belden, Phil
  background 35, 36
  boundary proposals 173
  Campaign, joins 40–1
  Campaign activities 61, 73, 74, 176
  television coverage 148
Bellamy, David 73

Benjamin, Floella 159
Benn, Hilary *11*, 141, 159, 160–1, 162, *167*, 169–72, 173, 175
Bentley Nib 131, 150
Birling Gap 28
Blackburn, Alan Chesters, Bishop of 86
Blair, Tony 61
Blair government 63–5
Blessed, Brian 134, *134*, 142, 159, 175
Bonington, Chris 73, 83
Bottomley, Sir Peter 159
boundary of the national park
  Campaign proposals 95, 102–4, 108–9, 111–13, 123
  character 45, 92
  Countryside Agency Area of Search proposals 99, 101, 104–5, 111, 122–3, 165
  Public Inquiry (2003–2004) discussions 114, 118, 119–23
  Public Inquiry (2003–2004) proposal 131
  Public Inquiry (2008) consultation on additional areas 165–6, 168
  Public Inquiry (2008) proposal 153–6
  Western Weald, change of outcome 164–5
Brandon, Peter 43, 54, 116
Brighton 62, 63, 166, 168
Brighton Borough Council 48
Brighton County Borough Council 28–9
Brighton & Hove City Council 29, 48, 57, 72, 166, 168
Bryant, Peter 84
Bryson, Bill 138–9, 157–8, 159, *167*, 175
Burney, Karen 161–2

Caborn, Richard 82
Cameron, Ewen 86, 87, 89
Campaign *see* South Downs Campaign
Campaign for National Parks (CNP) 41–2, 76–8, 100, 120, 124, 125
Campbell, William 28
Carden, Sir Herbert 28–9
Carden, John 136, 175

Carnarvon, Henry Herbert, 7th Earl of 97
Cartwright, Elizabeth 123, 135
Castle Goring 166
celebrities *see* national figures
chalk downlands 40
chalk grasslands 16–17, 36, 38
chalk rivers 21, *22*
'chalk-only' national park 51, 92–3, 119, 121, 122–3, 124–5, 131
Chambers, Rodney 151, 153
Chambers, Ruth 79, 95, 99–100, 132, 143, 148–9, 176
Chatsworth Hotel, Worthing 117, 148, 168
Cheater, Brian 152
Cheesman, Robert 139, 140, *144*, 149, 162
Cheltenham 110
Cherrett, Trevor 68
Chesters, Alan, Bishop of Blackburn 86
Chichester District Council 48, 71, 108, 117–18, 127, 136
Clark, Len 54, 73–4, 84, 135, 178
Clegg, David 175
Clegg, Rodney 73
conferences 39–40, 67–70
  see also political party conferences
conservation boards 46–7, 48, 107
  see also Sussex Downs Conservation Board
Conservative Party 48, 55
Council for National Parks 41–2
  see also National Parks Committee
Council for the Protection of Rural England (CPRE) 29, 50, 93, 94, 138
CPRE East Hampshire 94
CPRE Hampshire 102, 103, 152
CPRE Sussex 39, 151, 153
Country Landowners Association 83–4
Countryside Agency
  Area of Search 99, 101–5
  Board meeting (December 1999) 86–7
  boundary of SDNP 96
  creation of 80
  Designation Order 110–11
  designation proposals 88–90
  designation timetable 91–2, 106
  Local Authority Consultation 106, 107–9
  planning responsibilities of national parks 96–7, 99–100
  Public Inquiry (2003–2004) 114–27
  requested to consult on SDNP 83, 84–6
  stakeholder meetings 94, 99
  see also Countryside Commission; Natural England
Countryside and Rights of Way Bill 91–2, 96, 97
Countryside Commission
  and AONBs 35, 36
  consultation on AONBs 66–7, 70–3
  national park management 35
  Position Statement (1992) 55–6
  Protecting our Finest Countryside (1998) 75
  recommendations for the South Downs 73–4
  SDNP, rejection of 75
  South Downs Conference (1997) 67–70
  and Sussex Downs Conservation Board 46, 47, 50–1, 56, 59, 64–5
  see also Countryside Agency
Countryside Stewardship Scheme (CSS) 76
Crane, Robin
  biodiversity study 116
  Campaign closing address 174–7
  Campaign discussion paper 90
  Confirmation Order *11, 167*
  Confirmation Order, signing of and celebrations 169–70, 172
  Countryside Agency Board meetings 87, 89
  Countryside Agency Governance Group 94, 100
  CPRE meeting (2000) 93
  Department of the Environment, Transport and the Regions meetings 74, 78
  environmental schemes 125–6
  Guardian column on announcement of the SDNP 163–4
  Kirdford and Plaistow 108
  letters of 48, 53, 86
  Meyrick Case 129–30
  national figures visit to Older Hill 142
  Public Inquiry (2008) closing statement 156
  roles of 40, 41, 43, 44, 178

South Downs Conference (1997) 69–70
Western Weald 136
criteria for designation 30, 78, 128

Defra *see* Department for Environment, Food and Rural Affairs
Denton-Thompson, Merrick 68
Department for Environment, Food and Rural Affairs (Defra) 107, 129, 133, 136, 141, 159, 160–1
Department of the Environment 52
Department of the Environment, Transport and the Regions 64, 74, 77
designation process timetable 91–2, 106
development management 124
    see also planning
Devil's Dyke 29, 36
Dibden, Caroline 130
Dimbleby, David 138, 159, 175
district councils 92–3
Ditcham Barn 138
Ditchling 108–9, 111, 131, 133, 140–1, 149, 162, 169, 171
Ditchling Society 140–1
Doughty, Sir Martin 109
Dower, John 30, 32, 39–40
Dower, Michael 56
Dower, Pauline 40
Dower Report 30, 32

Eagle, Angela 64, 65, 66
East Hampshire AONB 33, 55, 92, 101, 169, 170
    see also South Downs Joint Committee
East Hampshire Association of Town and Parish Councils 94
East Hampshire District Council 55, 92–3, 94, 123, 135, 152
East Sussex County Council 29, 35, 51–2, 118, 140
Eastbourne Borough Council 28, 48
Eastern Sussex Downs ESA 38
Edwards, Penny 40, 41
Edwards, Ron 39, 87
Edwards, Victoria 87
Edwards Review 39, 43
    '*Fit for the Future*' 48–50, 73

Elcoate, Vicki 76–7, 78, 79, 81–2, 83, 86, 90
Ellerby, Jack 140
Elliott, Ian 44, 68, 74
Elsted 161–2
English Nature 62, 109
Environment Act (1995) 57
Environmentally Sensitive Areas (ESAs) 37–8, 62, 76
    see also subsidies
estates 24

Fairbrother, Chris 146, 161, 169, 170
Farmer, Alison 155
farming *see* agriculture; ploughing
Fewster, Jacquetta 146, *147*, 152, 153, 172
Flight, Howard 64, 118
Fogle, Ben 142, 159, 175
Friends of the Earth 58, 69, 77, 79
Friends of Lewes 139, 140

General Elections 61, 63
geology 14, *15*, 17
governance 100
Graffham Down 36–7
grassland *see* chalk grasslands
Green Ridge 166, 168
Griffiths, Robert 118, 125, 126–7, 143, 153, 154, 156–7
Guardian 47, 163–4
Gummer, John 63

Halstead, Sue 152
Hampshire County Council
    Area of Search proposals 123
    Countryside and Rights of Way Bill 96, 97
    East Hampshire AONB 33, 55
    Landscape Assessor's reports 150
    SDNP, support for 92
    South Downs Conference (1997) 68
    Western Weald proposals 135, 143, 145, 146, 148, 149, 156, 165
Hampshire Hangers 17, *18*, 101, 131
Hampshire Ramblers 93
Hampshire Wildlife Trust 83, 93
Hankinson, Moira 117, 122, 154–5
Hanson, Derek 73
Harting Down 17, 133–4, *134*

Harvey, David 41
Harwood, Stephen 93, 94
heathlands 19, *20*, 21
Hendon, Nicola 147–8
Herbert, Nick 160
Heritage Coast 35, 173
Hill, Ian 125
Hinds, Damien 143
Hinton Estate 128
Hobhouse, Sir Arthur 30, 34
Hobhouse Committee 30–2, 34
Hobhouse Report 60th anniversary 133–4, *134*
Hove Borough Council 48
  see also Brighton & Hove City Council
Hughes, Colin 136, 172
Huhne, Chris 175
Hunt, Jeremy 135
Hunt, John, Lord 42

Inspector *see* Parry, Robert Neil
Irranca-Davies, Huw 159, 160, 175
Itchen, river 21, 23, 101

Johnson, Sir John 52, 54
Johnson-Smith, Sir Geoffrey 78
Johnston, Norma 73
Judd, Frank, Lord 175
judicial reviews 128

Keep Our Downs Public 57–8
Keep The Ridge Green 166
Kimball, Marcus, Lord 80–1
Kirdford 108, 111, 131
Kleinwort Trust 100

Labour Party 48, 58, 63, 83
Landscape Assessor 119, 131, 141, 145, 150, 151, 152–3, 154, 155, 156
Landscape Design Associates 118
landscape detractors 151
landscape experts 117, 118
landscapes 14, 40, 120–1
Leahy, Martin 118
Leonard, Pat 56, 74, 78, 87, 93, 125
Lepper, David
  Brighton & Hove City Council

consultation 72
Green Ridge campaign 166, 168
House of Commons debates 78, 97, 98
Offham Down SSSI and 63
petitions 79
postcards to Defra 159
South Downs Conservation Board 64
support of 174–5
Lewes 25–6, *27*, 104, 121, 131, 133, 139–40, 149, 162, 172
Lewes District Council 140
Lewes Town Council 140
Lewis, Rhodri Price 117, 118, 127
Lewis, Sophie-May 137
Liberal Democrats 48
Liss 33, 92, 93, 101, 102, 131, 135, 151–2
Liss Parish Council 151
Loat, Emma 79, 83, 95
Local Government Commission 56
Lodge Hill 162
Lomas, John 125
Longmoor 109, 138, 152
Loughton, Tim 78, 140
Low Weald 101, 119
Lyall, Andrew 44, 78, 176
Lynees, Scott 117

Mann, Andrea *147*
Manpower Services Commission 36
Marine Conservation Zones 173
market towns *see* towns
markets 24–5
Marrington, Emma 138, 151
Marshall, Alison 73, 74
Mates, Michael 135, 137, 159, 175
McDonald, Ramsay 29
Meacher, Michael 172, 174
  briefed badly 79
  Campaign response to Countryside Commission 75–6
  consultation on AONBs 66–7
  and Offham SSSIs 63
  request for information 77–8
  request to Countryside Agency for South Downs review 84–6
  support for national park 82, 88
Sussex Downs Conservation Board

extension 64
Meale, Alan 77, 78, 79
media coverage 177
   announcement of the SDNP 162, 163–4
   Bill Bryson and 139
   Campaign for SDNP 52, 73
   of damage to environmental schemes 62, 63
   effects of 58
   letters from the public 137
   misinformation from West Sussex County Council 59
   postcards to Defra 159
   Public Inquiry 148
   support for SDNP 127
Mersey, Richard Bigham, 4th Viscount 65
Meyrick Case 128, 129, 149
Midhurst 25, 92, 131, 172
Midhurst Town Council 136
military training 26
Millmore, Paul
   background 35–6
   boundary proposals 95, 112, *112*, 172–3
   Lewes, campaign for inclusion 140, *144*, 149, 162
   National Park, commitment to 41, 176
   Public Inquiry opening statement 115
   South Downs Joint Committee 130
Milner-Gulland, Robin 118
Ministry of Defence 109, 138, 152
Moore, Kathy 142
Morris, Brian, Lord 80
Mullenger, Roger 153
Mullin, Chris 84, 96, 110
museum 24, 160

Napier, Christopher 130, 135, 136–7, 143, 149, 156
Nathan, Roger, 2nd Baron 59
national figures 134, *134*, 138, 142, 159, 175
National Park Authorities Review 107
national parks
   committees 34–5
   designation process 87–8, 91–2, 106
   development management 124
   establishment of 29–32, 38
   funding 35
   misinformation about 71, 72
   planning authorities 70–1
   political parties 48, 55
   purposes of 57
   reviews *see* Edwards Review
   see also Campaign for National Parks; Council for National Parks
National Parks and Access to the Countryside Act (1949) 32, 34, 43
National Parks Commission 32, 34, 35, 40
National Parks Committee 30–2, 41
   see also Council for National Parks
National Parks Review Panel 39
National Trust 28, 29, 83
Natural England 130, 134, 143, 146, 150, 153, 155, 156
   see also Countryside Agency
Natural Environment and Rural Communities (NERC) Act (2006) 129, 149
New Erringham Farm, West Sussex 62
New Forest National Park 116, 119–20, 122
New Forest National Park (Designation Order) 128
Newbould, Frank, *Your Britain fight for it now* 26, *27*
newspaper coverage 52, 59, 62, 63, 73, 127, 137, 139, 159, 163–4
Nobbs, Amanda
   background 42
   Campaign, joins 41
   Campaign activities 43, 50, 52, 74, 77, 79, 176
   speeches 39, 93, 94
noise 151–2
Nugent, Alastair 68

oak trees 161
O'Brien, Donna 120
Offham Down, East Sussex 62–3, *62*
Older Hill 142

Palmer, Minette 93, 129, 130
Paren, Margaret
   announcement of the SDNP 161–2
   boundary group 112, *112*, 113
   Campaign, joins 102
   Cheltenham, trip to 110

contribution of 176, 177
national figures and 138, 142, *167*
other activities 104, 109, 116, 129, 132
postcards to Defra 159, *167*
Public Inquiry (2008) closing statement 156
Public Inquiry (2008) submissions to 152, 153
South Downs Joint Committee 130
unveiling the plaque *11*
Western Weald campaign 135, 145–7, *147*, 148, 150
Paren, Nigel 151–2, 175
parish boundaries 23
Parry, Robert Neil (Inspector)
    Bill Bryson and 139
    boundary discussions 118, 149–50, 153, 155, 162
    character 114
    closing date for 2003–2004 Public Inquiry 126
    consultation on additional areas 168
    Public Inquiry (2008) approach to 143, 148
    Public Inquiry reports 131, 164
Patten, Chris 39, 42
Payne, Denis 175
Peacehaven 28
Perkins, Ben 60, 67, 152
Petersfield 24–5, 33, 92, 93, 101, 102–3, 121, 131, 135, 151
Petersfield Society 151
petitions 79, 136–7
Petworth 25, 92, 131
Phillips, Adrian 45, 50
Plaistow 108, 111, 131
planning
    Campaign views of 104
    Countryside Agency Technical Group 99–100
    local authorities' views 94–5
    in national parks 34, 69–71, 82, 96–7
    Public Inquiry (2003–2004) discussions 123–5
    Sussex Downs Conservation Board role in 68, 80
ploughing 62

Plumpton 166, 168
Plunkett, Owen 93, 102, 110, 112, *112*, 133, *167*
poem 137
political parties 48, 55, 58, 63
political party conferences 83, 109–10
Portslade race track 29
postcards to Defra 159, *167*
posters 147
power lines 151
prehistoric period 16, 19
Prescott, John 83
Private Members Bill (1999) 80–1
Public Inquiry (2003–2004) 114–27
    agri-environmental schemes 125–6
    boundary discussions 114–23
    consultation on 133
    Landscape Assessor's report, late submission of 141
    other administrative matters 126
    planning 123–5
    press release 127
    public hearings 126
    re-opening of 142
    report 130–2
Public Inquiry (re-opened 2008) 143–59
    Campaign closing statement 156
    consultation on additional areas 165–6, 168
    consultation on NPA membership 168
    NERC Act amendments, implications of 149
    pre-Inquiry meeting (2007) 143, 145–6
    report 164
    Western Weald and boundary discussions 148–56
public support 57, 78, 81, 143, *144*, 159, 177
Purves, Libby 175

Ramblers Association 50, 78, 79, 81, 109–10, 124, 125
recreation 45, 53, 76, 120–1, 152
    see also tourism
Redhead, Brian 42
Reed, Richard 44, 51, 142, 145, 151
Renton, Tim, Lord 65, 68, 69, 82, 97, 130, 140

Private Members Bill (1999) 80–1
Review of English National Park Authorities 107
Reynolds, Fiona 45, 52
Richmond, Emily 109–10
Ridley, Nicholas 42
rivers 21, *22*, 23
roads 38, 151
    see also A3 corridor
Rooker, Jeff, Lord 110
Rose, Francis 33, 102, 116, 175
Rother, river 23, 131, 151, 152–3, 155
Rowe, Frances 86, 89

Sandford Principle 57, 72
Schaffer, Sheila 175
Second World War 26, *27*
Selborne 19, 31, 32
Selborne, John Palmer, 4th Earl of 129
Seven Sisters *13*, 28
Shaw, Jonathan 134, 142
Shaw, Richard 169
Shaxson, Andrew 136, 151, 153, 172
Shoard, Marion 36–7, 38, 39–40
Short, Brian 41, 54
Short, Clare 110
Silkin, Lewis 172
Sites of Special Scientific Interest (SSSIs) 62–3, *62*
Smith, Henry 136, 147–8
Smith, Terry 95
Society of Sussex Downsmen 28, 29, 43, 60, 65, 100
    see also South Downs Society
Sompting Waste Management Complex 82
South Downs Advisory Forum 130
South Downs Campaign
    achievements of 88, 105, 170
    aims of 44–5, 46
    AONBs consultation, responses to 70–1, 72, 73, 74
    Area of Search proposals 102–4
    arguments for a national park 45–6, 51–2, 53–4, 61, 69
    boundary of SDNP 95, 108–9, 111–13, 115–16, 119, 120–1, 123
    Campaign officer 100
    closing of 173–4
    Confirmation Order 169–70
    Countryside Agency, views on 87
    creation of 40–1
    designation process timetable 91–2, 106
    funding 100, 175
    Local Authority Consultation 108, 109
    meetings attended (2000–2001) 93–4, 96, 99–100
    members of 43–4, 50, 59, 60, 180–4
    Meyrick Case 129
    ministers, pressure put upon 76–80
    planning group 95
    planning proposals 125
    points of principle 90
    proposals for national park 47, 50, 51, 63–4, 65
    Protecting our Finest Countryside (1998), response to 75–6
    Public Inquiry (2003–2004)
        report response 131–2, 141
        submissions 115–16, 119, 120–1, 123, 125–6
    Public Inquiry (2008)
        closing statement 156
        consultation on additional areas 165–6
        consultation on NPA membership 168
        pre-Inquiry meeting (2007) 143, 145–6
        preparations 146–8
        report response 166
        submissions 149, 150–4
    60th anniversary of the Hobhouse Report 133–4, *134*
    South Downs Conference (1997) 69
    South Downs Joint Committee members 130
    The South Downs: Securing their future (1997) 61
    Sussex Downs Conservation Board, views on 53, 58
    towns, support for 140–1
    weaknesses of and threats to 90
    Western Weald, fight for inclusion of 133–9, *134*
South Downs Campaign Group *see* South Downs Campaign
South Downs Conference (1997) 67–70

South Downs Integrated Landscape Character Assessment 150
South Downs Joint Committee 130
South Downs National Park
  announcement of the SDNP 160–4
  area of 14 *see also* boundary of the national park
  arguments against 45, 61–2, 75, 76, 104, 117, 118
  arguments for 29, 32, 40, 45–6, 52, 53–4, 61, 69, 78, 115–16
  Authority membership 168, 173, 177
  character 14–26, *15*, *18*, *20*, *22*, *27*, 31
  Confirmation Order *167*, 169–70, 171
South Downs National Park Campaign *see* South Downs Campaign
South Downs Network 174
South Downs Society 118, 130
  see also Society of Sussex Downsmen
Standing Committee on National Parks 30
  see also National Parks Committee
Steyning 140
Storrington 113
Streeter, David 45
Struthers, Tony 151
subsidies 62, 125–6
  see also Environmentally Sensitive Areas (ESAs)
Sullivan, Mr Justice 128
surveys 31, 32, 76, 78, 111–12, 152
Sussex Archaeological Society 50
Sussex Downs AONB 33, 46–7, 92, 169, 170
Sussex Downs AONB Forum 36, 47, 50
Sussex Downs Conservation Board
  arguments against a national park 74
  arguments against including the Weald 121–2
  consultation on future of 58–9
  creation of 50–1, 54
  deficiencies of 53
  planning process, weakness in 80, 82
  response to *The South Downs: Securing their future* (1997) 61–2
  South Downs Conference (1997) 68
  successes of 64–5, 76
  support for national park 84
  threats to 56
  see also South Downs Joint Committee
Sussex Rural Community Council 39, 50, 84
Sussex Wildlife Trust 40–1, 44, 100
Symonds, Richard 59

Tagtallia-Kershaw, Linda 146
Tansley, Arthur 31, 33
television coverage 52, 148, 159, 162
Templeton, John
  Area of Search proposals 102–3
  background 59
  boundary proposals 95, 112, *112*, 113, 165
  Confirmation Order *167*
  Countryside Agency Board meetings 89, 110
  Meyrick Case 129
Theobald, Geoffrey 72
Thornber, Ken 97, 123
timetable for designation process 91–2, 106
Todd, Chris
  announcement of the SDNP 161
  appointed as Campaign Officer 100–1
  Bill Bryson and 138
  boundary working group 112, 113
  called for resignation of Lord Renton 82
  Confirmation Order 170
  contribution of 176–7
  joined Campaign 67
  Local Authority Consultation 108, 109
  Marine Conservation Zones 173
  Meyrick Case 129
  newsletters 104
  Offham Down SSSI and 62–3
  postcards to Defra 159
  preparation for public consultation 102–3
  Public Inquiry (2003–2004) report, response to 132, 141, 142
  Public Inquiry (2008) preparations 146, 147, 148, 150
  South Downs Conference (1997) 69
  summary of Campaign achievements 105
  unveiling the plaque *11*
  work of 61, 77, 79
tourism 71
towns 24–6, 121, 131, 133, 151
train noise 151–2
Tregay, Robert 99, 118, 145, 146, 153–4, 155

Turner, Des 78, 166, 168, 174–5
Tyrie, Andrew 78, 107, 118, 173

Venning, John 136, 138, 145, 153, 175
villages 24
visitor pressure 71

walks and walking 133–4, *134*, 137, 138, 152–3
Waterson, Nigel 98, 118
Weald and Downland Living Museum 24, 160
Weeks, Honeysuckle 138, 175
West Sussex County Council
  announcement of the SDNP 162–3
  AONB review 46, 47
  arguments against a national park 71
  inaccurate information from 59, 72, 84, 109, 136
  judgement of, clouded 127
  Kidford and Plaistow, opposed to inclusion of 108
  needs of 51–2
  planning concerns 94–5
  Public Inquiry (2003–2004), submissions to 117–18, 121
  Sompting Waste Management Complex 82
  South Downs Conference (1997) 68
  Steyning, opposed to inclusion of 140
  Western Weald, opposed to inclusion of 141–2, 147–8
  Western Weald, shift of position on 148–9
West Sussex Gazette 127
Western Weald
  arguments against inclusion 121–2, 148
  arguments for inclusion 33, 51, 135, 157–8
  change of outcome 164–5
  character 13, 14, 19
  fight for inclusion of 133–9, 141–2
  oak trees 161–2
  planning, difficulties with 125
  pre-Inquiry meeting (2007) 143, 145–6
  proposed exclusion of 131
  Public Inquiry (2008) submissions to 149–56
  towns and villages 92, 108

Whitty, Larry, Lord 97
Williams, Rendel 116
Winchester City Council 92
Wood, Victoria 89
woodlands 17, *18*, *20*, 21, 23
Woolmer Forest 19, 109, 111, 152
World War II 26, *27*
Worthing Society 166
Worvell, Kath 175

AUTHOR ROBIN CRANE

After seven years as a regular army officer Robin Crane joined Guinness as a maltster. This led him to Lincolnshire where he became actively involved with the Lincolnshire Wildlife Trust. A film he made for the Trust resulted in the BBC Natural History Unit recruiting him to make programmes for the "World About Us" series.

He moved to Sussex in 1967 where he established his own production company and made environmental, scientific, equestrian, training and public relations films which won several national and international awards.

He became chairman of the Sussex Wildlife Trust and then chairman of the Royal Society for Nature Conservation In 1999 he was awarded the CBE for services to nature conservation.

Meanwhile in 1990 he chaired the first meeting of the South Downs Campaign. He remained as chairman, treasurer and fund-raiser until the South Downs National Park was confirmed. In recognition of this work he was awarded an Honorary Doctorate of the University of Sussex in 2010.

Editor MARGARET PAREN

Margaret Paren. a retired senior civil servant, has kindly edited the book as well as writing most of the accounts of the Public Inquiry. She was awarded an OBE for her work as vice-chairman of the South Downs Campaign in its later years and was chair of the South Downs National Park Authority for its first ten years